Decentralised Wastewater Treatment Systems (DEWATS) and Sanitation in Developing Countries

A Practical Guide

Editors: Andreas Ulrich, Stefan Reuter and Bernd Gutterer

Authors: Bernd Gutterer, Ludwig Sasse, Thilo Panzerbieter and Thorsten Reckerzügel

Water, Engineering and Development Centre,
Loughborough University,
Leicestershire, LE11 3TU, UK

© BORDA, 2009
Designed and produced by Bremen Overseas Research and Development Association (BORDA), Germany Phone: +49 (0) 421 137 18 Fax: +49 (0) 421 165 53 23
E-mail: office@borda.de

ISBN: 978 1 84380 128 3

All rights reserved. No part of this publication may be reprinted or reproduced or utilized in any form or by any electronic, mechanical, or other means, now known or hereafter invented, including photocopying and recording, or in any information storage or retrieval system, without the written permission of the publishers.

A catalogue record for this book is available from the British Library.

WEDC (The Water, Engineering and Development Centre) at Loughborough University in the UK is one of the world's leading institutions concerned with education, training, research and consultancy for the planning, provision and management of physical infrastructure for development in low- and middleincome countries.

This edition is reprinted and distributed by Practical Action Publishing.
Since 1974, Practical Action Publishing has published and disseminated books and information in support of international development work throughout the world. Practical Action Publishing trades only in support of its parent charity objectives and any profits are covenanted back to Practical Action (Charity Reg. No. 247257, Group VAT Registration No. 880 9924 76).

All reasonable precautions have been taken by the WEDC, Loughborough University to verify the information contained in this publication. However, WEDC, Loughborough University does not necessarily endorse the technologies presented in this document. The published material is being distributed without warranty of any kind, either expressed or implied. The responsibility for the interpretation and use of the material lies with the reader. In no event shall the WEDC, Loughborough University be liable for damages as a result of their use.

BORDA was founded in 1977 in Bremen Germany as a non-profit professional organisation with the goal of developing new methods of using renewable energy to alleviate poverty and, through the implementation of development programmes, to improve the living conditions and social structures in disadvantaged communities abroad.

Unlike other organisations, in the struggle against poverty BORDA focuses on the facilitation of basic needs services in the sectors of water, wastewater, solid waste and energy. To achieve this, partner structures, with the participation of all stakeholders, are advised and assisted in the stablishment and organisation of innovative basic needs services (BNS); this occurs during all phases of planning and construction up to the stages of operation and maintenance.

Editors of the publication are:
- Andreas Ulrich (BORDA Director)
- Stefan Reuter (BORDA Vice Director)
- Bernd Gutterer (PhD, International Consultant)

Authors of the publication are:
- Bernd Gutterer (PhD):   Chapters 1–7
- Ludwig Sasse: Chapters 7–10, Sections 11.4–11.5
- Thilo Panzerbieter   Sections 11.1–11.3
  (Section 11.3 in collaboration with Andreas Schmidt)

Thorsten Reckerzügl has provided substantial documentation.
Editorial contributions: Mary Breen and Michael Smith

Acknowledgements

This publication is a collective effort. Since the early 1990s BORDA has collaborated with a multitude of individuals and institutions throughout Europe and Asia to develop the DEWATS approach. The first DEWATS Handbook was published by Ludwig Sasse in 1998. It served as an instruction manual focusing on the technical design. A wealth of experience in demand-oriented technology adaptation and dissemination has evolved since then, including public health and community-based sanitation. This book presents the collaborative efforts made by a wide range of professionals from local and central authorities, from private businesses and international donors, NGOs, community-based organisations and academia. Therefore, this publication could not have been realised without the generous contribution of the many individuals and organisations who shared their experience and expertise. In particular the editors would like to express their gratitude to following partner organisations and individuals:

- Indonesia: LPTP (Surakarta); BEST (Tangerang); BaliFokus (Denpasar)
- India: Consortium for DEWATS Dissemination Society (CDD, Bangalore) and the associated partner network
- China: Sustainable Development Strategy Institute (SDSI) at Zhejiang University of Technology (ZUT), Hangzhou
- BORDA's Regional Programme Co-ordinators Frank Fladerer (BORDA – South-East Asia), Pedro Kraemer (BORDA – South-Asia) and Andreas Schmidt (BORDA – Southern Africa)
- Prof. Chris Buckley (Pollution Research Group, University of KwaZulu-Natal, South Africa) and Ludwig Sasse (retired, pioneer of BORDA's Biogas and DEWATS solutions)

TABLE OF CONTENTS

1	Introduction	12
2	**Towards comprehensive wastewater and sanitation strategies**	**14**
2.1	World water resources under threat	14
2.2	The protection of water resources – achievements and challenges	17
2.3	A short assessment of the sanitation and wastewater sectors in developing countries	20
2.4	Signs of change – elements of efficient and sustainable sanitation programmes	27
2.5	Towards service orientation – the conceptual framework of basic needs sanitation programmes	29
2.6	The increasing demand for efficient and reliable decentralised wastewater-treatment solutions	31
3	**DEWATS – Sustainable treatment of wastewater at the local level**	**32**
3.1	DEWATS – a modular system approach to ensure efficient wastewater treatment performance	33
3.2	DEWATS – a brief insight into technical configuration	35
3.3	DEWATS – good practice examples/applications	37
3.3.1	DEWATS/CBS – Community-Based Sanitation programme in Alam Jaya, Tangerang, Java, Indonesia	37
3.3.2	DEWATS/CBS – Community-Based Sanitation programme in Ullalu Upanagara, Bangalore, India	40
3.3.3	DEWATS at public institutions – Sino-German College of Technology, Shanghai, China	43
3.3.4	DEWATS at public institutions – Aravind Eye Hospital in Thavalakuppam, Pondicherry, India	46
3.3.5	DEWATS/SME-Cluster approach – Kelempok Mekarsari Jaya small-scale industry cluster, Denpasar, Bali, Indonesia	49
3.3.6	DEWATS/SME – Alternative Food Process Private Ltd. Bangalore, Karnataka, India	52
3.3.7	Infrastructural development in rural China – Longtan Village, Danleng County, Szechuan Province, China	54
3.3.8	DEWATS in integrated municipal planning – Wenzhou University, Zheijang Province, China	56

4	**Mainstreaming DEWATS – strategic planning and implementation of sustainable infrastructure**	**58**
4.1	Strategic planning of sanitation programmes	58
4.2	Legal framework and efficient law enforcement	62
4.3	Target-oriented local and municipal planning	66
4.3.1	Features of urban infrastructure development	66
4.3.2	Sanitation mapping as a tool for efficient urban-infrastructure development	68
4.4	Financial analysis	74
4.4.1	Comparative cost analysis for infrastructure development	74
4.4.2	Economic analysis in times of global warming and energy scarcity	77
4.4.3	Economic considerations for point-source polluters	80
4.4.4	Parameters for economic calculation	82
4.4.5	Sustainable financing schemes for sanitation programmes – multi-source financing and willingness to pay	88
5	**CBS programme planning and implementation**	**92**
5.1	Stakeholders in CBS programmes	92
5.2	Responding to basic needs – active involvement of beneficiaries and residents	96
5.3	Local government and municipality bodies	97
5.4	Non-governmental organisations	99
5.5	Private sector	99
6	**CBS Programme – detailed procedure for implementation**	**100**
6.1	First planning activities	100
6.2	The pilot project	103
6.3	Preparation phase	105
6.3.1	Kick-off workshop	105
6.3.2	Planning workshop	106
6.3.3	Community pre-selection and community assessment	108
6.4	Planning phase	110
6.4.1	Site assessment	110
6.4.2	Informed technology choice	112
6.4.3	Detailed engineering design	114
6.4.4	Economic planning	116
6.4.5	Agreement on implementation and landholding	117

6.5	Implementation phase	118
6.5.1	Task planning	118
6.5.2	Quality management	120
6.5.3	Construction	121
6.5.4	Pre-commissioning test	123
6.5.5	Parallel training measures	123
6.6	Operation phase	124
6.6.1	Start operation	124
6.6.2	Operation & maintenance	126
6.6.3	Use of biogas	129
6.6.4	Monitoring and evaluation	131
7	**DEWATS components & design principles**	**132**
7.1	Basics of wastewater treatment	132
7.1.1	Definitions: pollution & treatment	132
7.1.2	Biological treatment	133
7.1.3	Aerobic – anaerobic	135
7.1.4	Physical treatment processes	136
7.1.5	Elimination of pollutants	141
7.1.6	Ecology and self-purification in nature	145
8	**Treatment in DEWATS**	**150**
8.1	Parameters for wastewater-treatment design	152
8.1.1	Control parameters	153
8.1.2	Dimensioning parameters	163
9	**Technical components**	**168**
9.1	Overview of DEWATS components	168
9.2	DEWATS modules	176
9.2.1	Grease trap and grit chamber	176
9.2.2	Septic tank	177
9.2.3	Fully mixed digester	182
9.2.4	Imhoff tank	184
9.2.5	Anaerobic baffled reactor	187
9.2.6	Anaerobic filter	191

9.2.7	Planted soil filters	195
9.2.7.1	Horizontal gravel filter	197
9.2.7.2	Vertical sand filter	207
9.2.8	Ponds	211
9.2.8.1	Anaerobic ponds	212
9.2.8.2	Aerobic ponds	216
9.2.9	Hybrid and combined systems	221
9.3	Non-DEWATS technologies	223
9.3.1	UASB	223
9.3.2	Trickling filter	225
9.3.3	Aquatic-plant systems	228
10	**Designing DEWATS**	**230**
10.1	Technical spreadsheets – background	230
10.1.1	Usefulness of computer calculation	230
10.1.2	Risks of using simplified formulas	231
10.1.3	About the spreadsheets	233
10.2	Technical spreadsheets – application	236
10.2.1	Assumed COD/BOD ratio	236
10.2.2	Domestic wastewater quantity and quality	237
10.2.3	Septic tank	238
10.2.4	Fully mixed digester	241
10.2.5	Imhoff tank	247
10.2.6	Anaerobic baffled reactor	250
10.2.7	Anaerobic filter	255
10.2.8	Horizontal gravel filter	261
10.2.9	Anaerobic pond	264
10.2.10	Aerobic pond	270
10.3	Spreadsheets for costings	274
10.4	Using spreadsheets without a computer	278

11	Project Components: sanitation and wastewater treatment – technical options	282
11.1	Toilets	283
11.1.1	Common practices to be discouraged	284
11.1.2	Closed pit toilets	286
11.1.3	Composting toilets	289
11.1.4	Dry, urine-diversion toilets	290
11.1.5	Pour-flush toilets	292
11.1.6	Community toilet blocks	296
11.2	Collection systems	297
11.2.1	Rainwater drains	297
11.2.2	Conventional gravity sewerage	298
11.2.3	Simplified gravity sewerage	299
11.2.4	Vacuum sewerage	303
11.3	Sludge accumulation and treatment	306
11.3.1	Sludge removal	307
11.3.2	Sludge treatment	308
11.3.2.1	Small-scale application – drying and composting	309
11.3.2.2	Large-scale application – sludge and septage-treatment facility	313
11.4	Reuse of wastewater and sludge	318
11.4.1	Risks	318
11.4.2	Groundwater recharge	321
11.4.3	Fishponds	321
11.4.4	Irrigation	324
11.4.5	Reuse for process and domestic purposes	324
11.5	Biogas utilisation	325
11.5.1	Biogas	325
11.5.2	Scope of use	327
11.5.3	Gas collection and storage	328
11.5.4	Distribution of biogas	333
11.5.5	Gas appliances	334

12	**System malfunction – symptoms, problems, solutions**	**336**
12.1	Insufficient treatment of wastewater	336
12.2	Reduced flow at the outlet of the facility	344
12.3	Other problems and nuisances	349
13	**List of abbreviations**	**350**
14	**Appendix**	**352**
14.1	Geometric formulas	352
14.2	Energy requirement and cost of pumping	352
14.3	Sedimentation and flotation	353
14.4	Flow in partly filled round pipes	354
14.5	Conversion factors of US-units	355
15	**Bibliography**	**356**

Introduction

Water is a key feature of public concern worldwide. Inappropriate use and poor management of water resources have an increasingly negative effect on economic growth, on social welfare and on the world's eco-systems.

For a long time the need for efficient wastewater treatment was ignored by many public authorities. As a result the performance of existing treatment technologies and the conditions of sanitation facilities are rather poor. At many locations the sewage is just drained to surface or ground waters without adequate handling.

Recently, decision makers, planners, engineers and civil society stakeholders have launched multiple initiatives to answer the question facing many developing countries: *How to ensure a good performance and a high coverage of wastewater treatment under rather difficult conditions with financial constraints and limited human and institutional capacities?*

In the 1990s an international network of agencies and NGOs drew conclusions about the deficiencies of existing infrastructure development and produced the so-called "DEWATS approach." DEWATS is designed to be an element of comprehensive wastewater strategies: not only the technical requirements for the efficient treatment of wastewater at a given location, but the specific socio-economic conditions are also taken into consideration.

By its principles of "reliability" and "longevity", the permanent and continuous treatment of wastewater flows ranging from 1–1000m^3 per day, from both domestic and industrial sources, should be guaranteed. With its flexibility, efficiency and cost effectiveness, these systems are planned to be complementary to centralised wastewater treatment-technology and to strategies reducing the overall generation of wastewater.

The international discussion about the conservation of water resources and more target-oriented poverty-alleviation strategies create a favourable environment for new sanitation approaches and innovative wastewater treatment solutions. In many countries a rapidly upcoming market for DEWATS and a demand for efficient Community-Based Sanitation (CBS) can be observed.

Based on the experiences and "good practice" of numerous programmes and projects, this book aims to present the most important features for successful DEWATS dissemination:
- driving forces and decision parameters for innovative wastewater and sanitation strategies.
- options for a comprehensive technology choice
- planning instruments for wastewater treatment and sanitation mapping
- presentation of the DEWATS approach and good practices in DEWATS
- basic knowledge about the process of wastewater treatment
- the technical components of DEWATS
- design principles for DEWATS
- guidelines for programme development and implementation of DEWATS based CBS programmes.

Since wastewater treatment and sanitation, with all its implications, is such a complex subject, the content focuses on providing a basic knowledge that is relevant for DEWATS dissemination. As a practical guideline it should support decision making, planning and implementation activities. For very specific questions, additional literature can be consulted. A selection of books and articles can be found in the appendix.

Andreas Ulrich   Stefan Reuter   Dr. Bernd Gutterer

Bremen / Berlin November 2009

Towards comprehensive wastewater and sanitation strategies

2.1 World water resources under threat

Water stress occurs when the demand for water exceeds the available amount during a certain period or when poor quality restricts its use. Water stress causes deterioration of fresh water resources in terms of quantity (aquifer over-exploitation, dry rivers, etc.) and quality (eutrophication, organic matter pollution, saline intrusion, etc.). Source: European Environment Agency, EEA glossary, 2006

Water is the essential basis for all forms of life. Water is of utmost importance for human health and dignity. Water is crucial for sustainable social and economic development. However, world water resources are under threat. In the past 250 years the world has seen a tremendous increase both in population and economic activities. This development process has resulted in extensive social transformation and a rapidly increasing demand for natural resources. Urbanisation, industrial development and the extension of agricultural production have a significant impact on the quantity and quality of water resources. Overexploitation of water bodies and deterioration of water quality are global trends.

Today one-third of the world's population lives in countries suffering from moderate to high water stress.[1] Since the mid-1990s, some 80 countries, representing 40 per cent of the world's population, have been suffering from serious water shortages in urban and rural areas – in a lot of cases, the result of the socio-economic development over the recent decades.

World Commission on Water 1997

The increasing demand for freshwater sources and rapidly changing production and consumption patterns are directly linked with the pollution of ground and surface waters. "More than half of the world's major rivers are seriously depleted and polluted, degrading and poisoning the surrounding ecosystems, threatening the health and livelihoods of those who depend on them.[2]"

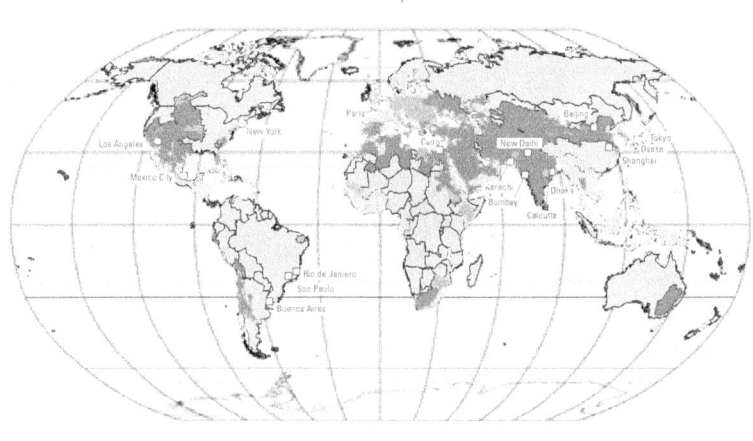

Figure 2_1:
Water stress (2000) in regions around megacities. This map is based on estimated water withdrawals for the year 2000, and water availability during the 'climate normal' period (1961-1990). Results shown in this map were calculated on river basin scale.
Source: WaterGAP2.1e by CESR, Kassel, Germany

withdrawal-to-availability ratio

- 0 – 0.2 low water stress
- 0.2 – 0.4 medium water stress
- more than 0.4 severe water stress

Picture 2_2:
More than half of the world's major rivers are seriously depleted and polluted

Although the threat to water resources is not only a phenomenon in developing countries, it is particularly the world's poor that are most affected: worldwide, 0.9 billion people still lack access to safe drinking water and 2.5 billion lack access to adequate sanitation. While improvements are monitored on the drinking water side, the challenge on the sanitation side obviously is much bigger than it was thought to be. Estimates indicate that approximately half the population of the developing world is exposed to polluted water resources, which increase disease incidence; most of these people live in Africa and Asia.[3]

3 JMP-WHO/Unicef 2008

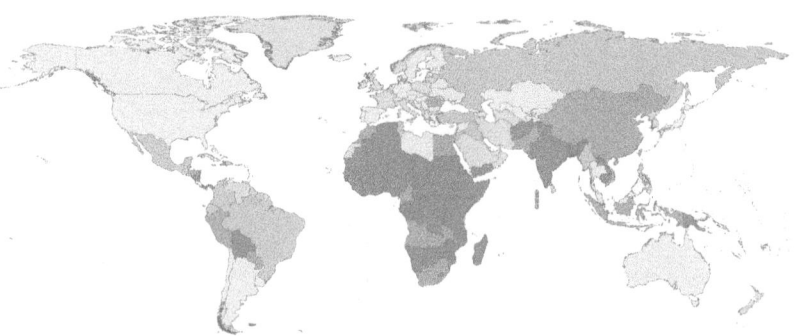

Percentage of population using improved sanitation

91% – 100%   76% – 90%   50% – 75%   Less than 50%   Insufficient date

Picture 2_3:
Half the developing world are still without improved sanitation;
Source: JMP-WHO UNICEF, 2008

Towards comprehensive wastewater and sanitation strategies

The challenges ahead are obvious: the urban population of the less-developed world is expected to nearly double in size between 2000 and 2030 from just under 2 billion to nearly 4 billion people, with the greatest urban growth occurring in Asia. By that time, 58 per cent of the world's population will live in urban or semi-urban areas. Pessimistic scenarios forecast that nearly 7 billion people in 60 countries will live in water-scarcity by 2050, whereas even rather optimistic projections estimate that just under 2 billion people in 48 countries will be affected.

Over the last 30 years, a multitude of national and multi-national initiatives have addressed the emerging water crisis. In 1980 the International Water Supply and Sanitation Decade (IWSSD) was launched. At the so-called Rio Conference in 1992 "water" was identified as one of the key elements for sustainable development:

Agenda 21,
Chapter 18

"The general objective is to make certain that adequate supplies of water of good quality are maintained for the entire population of this planet, while preserving the hydrological, biological and chemical functions of ecosystems, adapting human activities within the capacity limits of nature and combating vectors of water-related diseases."[4]

In September 2000, 189 UN member states adopted the so-called Millennium Development Goals, setting well-defined targets for the world's most pressing development issues. The seventh goal is to sustain the human environment. Its target number 10 is related to water supply and sanitation. Access to sanitation has been added at the Johannsburg Sustainable Development Summit.
Target 10 is:

Johannesburg Plan
of Implementation,
Paragraph 8

"To halve, by 2015, the proportion of people without sustainable access to safe drinking water and basic sanitation."[5]

In order to meet this sanitation target, an additional 1 billion urban dwellers and almost 900 million people in rural communities have to be served with adequate facilities by 2015; this equates to approximately half a million extra people to be serviced each day.

2.2 The protection of water resources – achievements and challenges

A number of encouraging results have been achieved by various initiatives launched on very different levels since the Eighties. While in 1990, only 77 per cent of the world's population used improved drinking-water sources, in 2002 a global coverage of 83 per cent was achieved. The deterioration of water-supply infrastructure in many developing countries – the per capita water-supplies had decreased by a third between 1970 and 1990 – was stopped in most places.

In the past few years, governments have developed more efficient approaches to halt the increasingly urgent water crisis:
- more efficient legal frameworks are being developed and, in many places, law enforcement has been improved
- water policy is increasingly recognised as a cross-cutting task for socio-economic development
- water resources are more comprehensively assessed through the application of new planning methods and technologies
- the conservation and sustainable use of water for food production and other economic activities receives more emphasis
- institutional and human capacity to assess and manage water resources are being created

Nonetheless, considering the number of people who are still without a safe water supply, the tasks required to meet the Millennium Development Goals are enormous.

In the fields of sanitation and wastewater treatment, the challenges are even greater. Although global sanitation coverage increased from 54 per cent in 1990 to 62 per cent in 2006, 2.5 billion humans still live without improved sanitation.[3]

Picture 2_4:
Slow progress in water and sanitati will hold back advances in other areas

Towards comprehensive wastewater and sanitation strategies

In India and China alone, nearly 1.3 billion people live without adequate facilities; in Sub-Saharan Africa, coverage extends to only 31 per cent of the population, and in Latin America and the Caribbean about 121 million people have no access – a critical situation.[3]

The impact of poor sanitation and water pollution is obvious:
In fact, some 1.5 million children die each year from preventable diarrhoeal diseases.[3]
In terms of disability adjusted life years (DALY), some 44 per cent of the annual global burden attributable to water, sanitation and hygiene are caused by diarrhoeal diseases. In other words, every year more than 52 million years of healthy life are lost (UN-WWDR3, 2009).[6]

DALY is a time-based measure that combines years of life lost due to premature mortality and years of life lost due to time lived in states of less than full health (WHO)

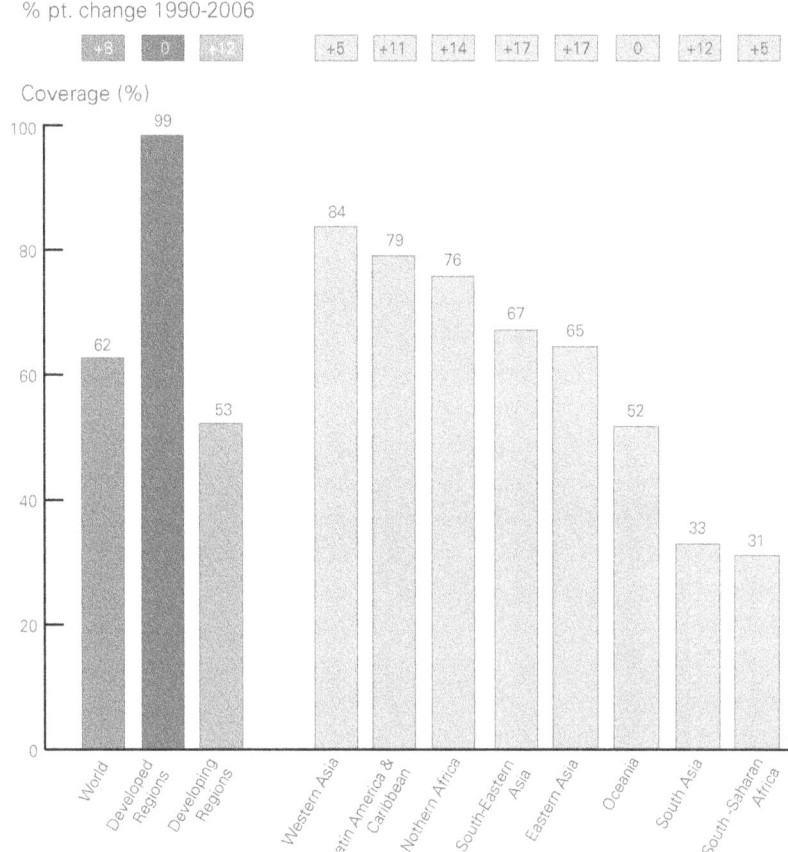

Figure 2_5:
Sanitation coverage is lowest in sub-Saharan Africa and South Asia.
JMP-WHO/UNICEF, 2008

It is the poor sections of these populations who are most affected by the increase of water-borne diseases. Many live in a vicious cycle of unhealthy living conditions, faecal-oral disease, illness and poverty.

The statistics on water borne disease underline the scale of the challenges that lie ahead:
- 443 million school days get lost each year from water-related illness
- one third of the world population (2 billion infections) affected by intestinal parasitic worms
- 6 million are blind from trachoma
- 200 million people are affected with schistosomiasis[7]

7 UNDP, HDR 2006

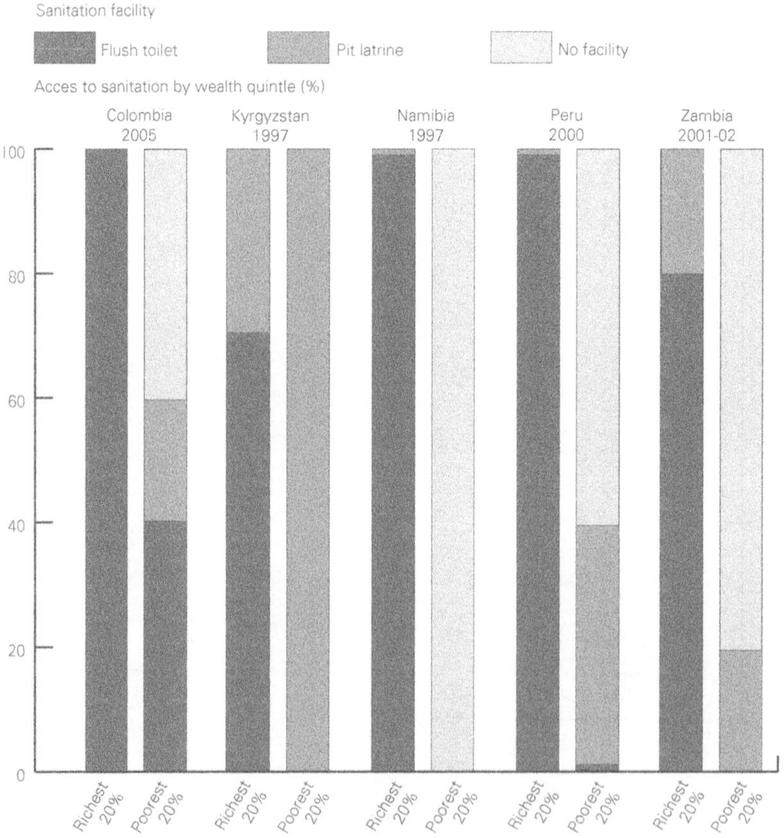

Picture 2_6:
Access to sanitation by wealth quintile selected countries
Source: UNDP, HDR 2006

2.3 A short assessment of the sanitation and wastewater sectors in developing countries

Poor sanitation should be perceived as an element of an overall process of inadequate use of water. Poorly treated wastewater has negative effects on public health. Furthermore, its high oxygen demand damages eco-systems, causing eutrophication in open water bodies, due to excessive nutrient supply; aquatic life is destroyed. Toxic substances also reach the groundwater, which approximately 2 billion people – about one-third of the world's population – depend on for water supply.

The three main sources of water pollution in developing countries are domestic, industrial and agricultural. The volume and characteristics of each type of wastewater differ by source and location (urban, rural). In total, domestic wastewater generally contributes the greatest organic load. In the Philippines, for instance, municipal (domestic) wastewater generates 48% of the national BOD (biochemical oxygen demand) (Industry 15%, Agriculture 37%); in Thailand, municipal wastewater generates about 54% of the total BOD.[8]

Water pollution from domestic, agricultural and industrial sectors results in tremendous public and private economic losses. Calculations of external costs indicate that in China about 2.6%, in Mexico about 3.3%, in India about 4.53%, in Eastern Europe up to 5% and in industrial countries between one and two per cent of the GDP is lost due to water pollution.[9] The World Bank estimates the annual losses to the Philippines' national economy to be about PhP 67 billion (US$ 1.3 billion); this can be broken down into PhP 3 billion for the health sector, PhP 17 billion for fishery production, and PhP 47 for tourism.[10] In Indonesia, economic losses are conservatively estimated at US$ 4.7 billion per year, which is roughly equivalent to US$ 12 per household per month.[11]

World Bank, 2005

CERNA, 2003

World Bank, Philippines Environment Monitor, 2003

World Bank, Indonesia Environment Monitor, 2003

Comprehensive sanitation strategies must therefore protect public health and the environment; they should include the collection, the treatment, the disposal, the recycling and, especially, the avoidance of waste. Wastewater strategies must address an array of different wastes:
- human excreta (urine + faeces = blackwater)
- household wastewater (shower + washwater = greywater)
- stormwater
- waste from industrial production
- hazardous waste, as from hospitals
- solid waste

Picture 2_7:
 Water pollution causes tremendo public and privat economic losses

Picture 2_8:
 Open drainage o untreated wastev is a prime source serious disease

For a long time, wastewater treatment systems in the "developed world" were seen as the ideal solution, which should also be applied in the "developing world". Wastewater treatment was perceived as a highly technical engineering task; flush toilets were used to transport the human excreta through big sewer systems to rather technically sophisticated wastewater-treatment plants.

However, a study carried out in 116 cities worldwide indicated a low connection of households to sewers in Africa, Asia, Latin America, the Caribbean and Oceania. Surveys show the rather weak efficiency of centralised wastewater-treatment systems.[12]

12 UN-WWDR1, 200

Towards comprehensive wastewater and sanitation strategies

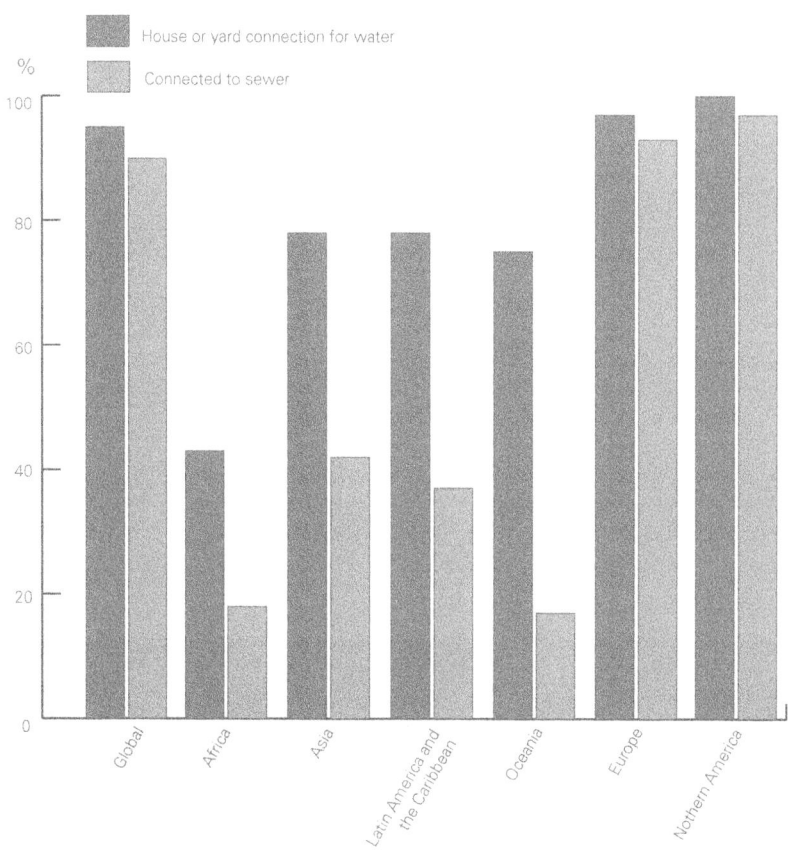

Figure 2_9:
The proportion of households in 116 major cities connected to piped water and sewers
Source: UN-WWDR1, 2003

Most of the sewage in developing countries is discharged to nature without adequate treatment. An assessment published by the Central Pollution Control Board in New Delhi indicates that only a small quantity of sewage flows to treatment plants in India. While in so-called "Class I cities" 33% of the collected and 24% of the total wastewater is treated, only 5.6% of the collected and 3.7% of the total wastewater is treated in smaller "Class II cities".

Class I cities are urban agglome-rations with a population of 100,000 or more, followed by: Class II (50,000 to 99,999), Class III (20,000 to 49,999), Class IV (10,000 to 19,999) and Class V (5,000 to 9,999).

Type	Number of cities/ towns	wastewater generated (MLD)	wastewater collected		wastewater treated		
			MLD	% (of generated)	MLD	% (of generated)	% (of total)
Class I Cities	299	16,662.5	11,938.2	72	4,037.2	33.8	24.0
Class II Towns	345	1,649.6	1,090.3	66	61.5	5.6	3.7
Total	644	18,312.1	13,028.5	71	4,098.7	31.5	22.4

Table 1: Sanitation coverage in India, Central Pollution Control Board, Delhi, 2005 1MLD = 1 million liters per day

These figures correlate with experiences in other countries. In the Philippines, only 7% of the total population is connected to sewers and more than 90% of the sewage generated in the Philippines is not disposed of or treated in an environmentally acceptable manner. Figures from Latin America and the Caribbean show that only 14% of the effluent is treated.[13]

13 Global Water Partnership, 2004

A closer look at the performance of existing wastewater-treatment systems reveals further reasons for the rapid deterioration of coastal waters and the dead waterbodies found in many countries. Technical and maintenance problems result in low treatment efficiency and a discharge of still highly contaminated effluent.

Picture 2_10: Power cuts and maintenance problems are frequently encountered at conventional and decentralised wastewater-treatment units

Poor performance is also observed in many so-called decentralised wastewater-treatment solutions, such as the:
- rotating disk reactor
- the trickling filter
- the activated sludge process
- the fluidised bed reactor and
- the sequencing batch reactor

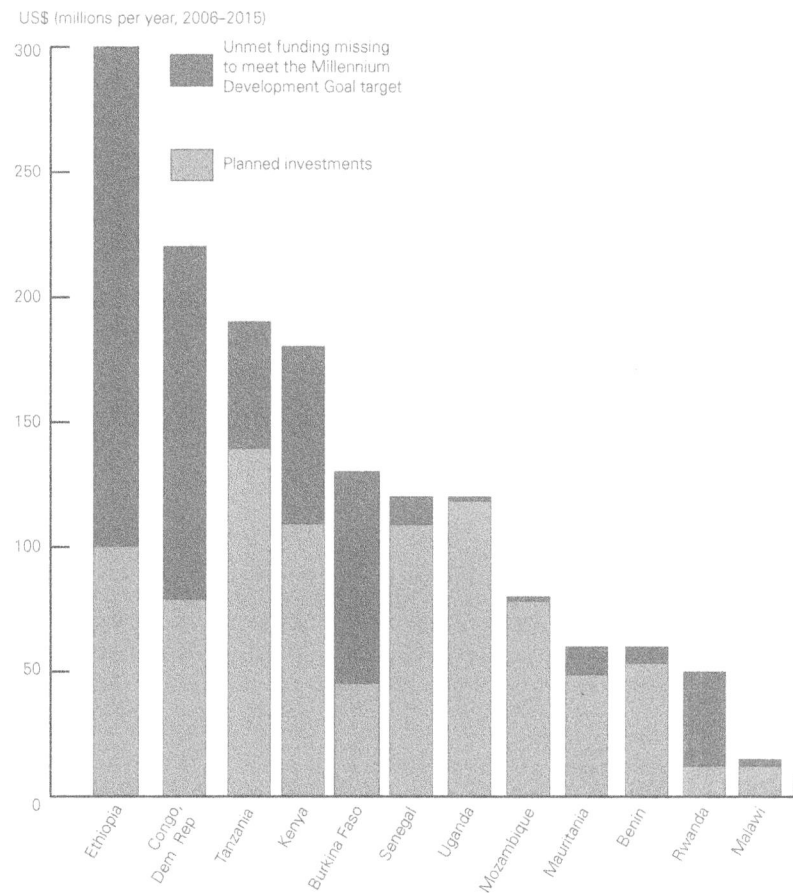

Picture 2_11:
The gap between planned investments and funding required to meet Millennium Development Goal in Sanitation;
Source: UNDP HDR 2006

The main causes for these treatment failures are insufficient operation and maintenance, lack of spare parts and frequent power-supply cuts.
A survey, carried out in 1999, showed that one-third of the centralised wastewater treatment plants in Thailand were malfunctioning or did not operate at all. Most of the facilities suffered from equipment failure or damage, as well as deficiencies in staff-skills levels. Performance data showed that – due to poor maintenance – the collection systems collected only 55% of the wastewater that the treatment plants were designed to treat.

Picture 2_12:
Conventional centralised systems quently suffer from poor performance

The increase of sanitation coverage has been slow because the extension of existing centralised systems has to be more complex than anticipated. In China, for instance, where increasing emphasis is placed on the treatment of the rapidly growing wastewater volumes, official reports state that the construction of about 700 – or half of the major wastewater-treatment projects planned by central government for the period 2001-2005 – had not yet been launched by the end of 2004[14].

The dive for such high sanitation standards is the result of a complex development process at local, regional and national levels. It includes elements such as:
- public awareness-raising and stakeholder involvement of civil society groups at all levels
- human and institutional capacity-building in engineering, private companies, science and public services
- application of relevant techniques and standards
- development of adequate legal frameworks and efficient law enforcement and
- availability and efficient allocation of financial resources.

14 State Environme Protection Agenc Administration o China (SEPA) ref. SINA 2005

Towards comprehensive wastewater and sanitation strategies

Due to the technological, institutional and organisational requirements, such complex wastewater projects are unthinkable in many parts of the world. There is not the availability of sufficient funds. Moreover, the high water demand for flushing toilets (30–50% of domestic water consumption) further increases water stress, particularly in arid and semi-arid regions.

The desperate need for the establishment and implementation of efficient sanitation programmes becomes even more apparent on viewing the slow progress between 1990 and 2006. The term "sanitation" frequently only refers to the collection, removal or disposal of human excreta. "Improved sanitation" facilities ensure hygienic separation of human excreta from human contact. Technical solutions are:

- flush or pour-flush toilet / latrine to:
  - piped sewer system
  - septic tank
  - pit latrine
- ventilated improved pit (VIP) latrine
- pit latrine with slab
- composting toilet

Although these technologies are fairly simple to implement and maintain, progress in implementing them remains rather slow in many countries.

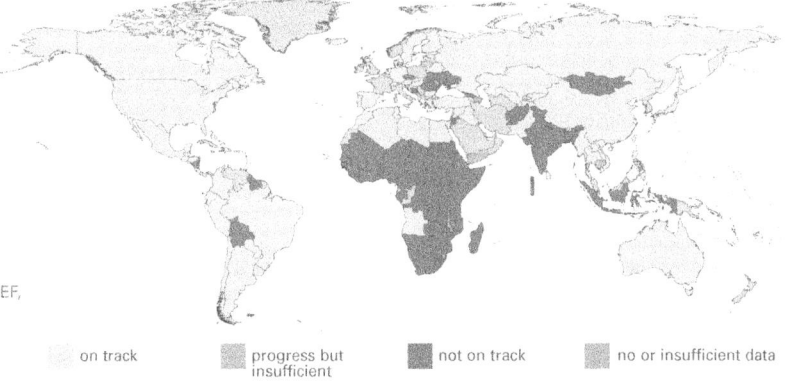

Figure 2_13: Progress in sanitation, 1990–2006
Source: JMP-WHO/UNICEF, 2008

on track | progress but insufficient | not on track | no or insufficient data

WHO estimates that meeting the target of the United Nations Millennium Development Goal for improved sanitation requires annual funds of approximately US$14 billion from the beginning of 2005 until the end of 2014[15].

WHO, Geneva 2008

2.4 Signs of change – elements of efficient and sustainable sanitation programmes

The reasons for slow progress in the sanitation sector are manifold. Performance both at the policy and implementation levels has been unquestionably weak in the past, resulting in unclear, contradictory or non-existent sanitation policies.

Typical political and administrative deficiencies are:
- lack of political will
- low prestige and recognition of the sector
- poor policy at all levels
- weak institutional framework
- inadequate and poorly used resources
- inappropriate approaches
- failure to recognise defects of current excreta-management systems
- neglect of consumer preferences
- ineffective promotion and low public awareness
- women and children considered last[16]

16 WHO, 1998

Within projects, considered at the implementation level, the following deficiencies have been observed:
- isolated character of the activities
- poor coordination between initiatives
- insufficient construction quality
- poor adaptation of designs to local conditions
- hardware-driven approaches
- insufficient involvement of users and other relevant local and regional stakeholders for implementation (top-down approach)

Recently, the situation has begun to change; national and international discussions are beginning to show results. In 2001, for instance, the Government of South Africa published a white paper on "Basic Household Sanitation". At that time, about 18 million citizens had no access to adequate
sanitation. Within the strategic paper, the government underlined its constitutional responsibility to ensure sanitation access to all South Africans.

Towards comprehensive wastewater and sanitation strategies

The purpose of the paper was to:
- spell out government policies on sanitation
- provide a basis for the formulation of local, provincial and national sanitation-improvement strategies
- provide a framework for municipal sanitation programmes
- ensure that sanitation-improvement programmes are adequately funded, and
- install mechanisms for monitoring the implementation of the policy and sanitation-improvement programmes[17]

Government of South Africa, 2001

Picture 2_14:
On the outskirts of major South African cities, people face poor sanitation

In 2002, the Government of Indonesia published a similar document, addressing the country's requirements for more efficient and sustainable implementation of safe water supply and sanitation. Within the paper, the government drew conclusions from problems faced in earlier programmes within the Water Supply and Environmental Sanitation (WSES) sector – and defined essential principles for future programmes:
- the role of all stakeholders involved in a programme must be clearly defined and their commitment ensured
- programmes must meet community's demand
- participatory management – involving all segments of the user community, especially women – is essential for successful long-term operation and maintenance
- the approach for environmental sanitation should be distinguished from that for clean water
- high quality of services is essential for meeting the expectations of the users and ensuring their willingness to pay for the services[18]

Ministry of Settlement and Regional Infrastructure, Ministry of Health, Ministry of Home Affairs, Ministry of Finance, National Development Planning Agency/ Bappenas, 2002

2.5 Towards service orientation – the conceptual framework of basic needs sanitation programmes

Although the documents discussed in the previous section were developed in a specific country context, both papers reflect ongoing, worldwide discussions concerning the development and implementation of successful sanitation programmes. Based on a broad range of position papers and experiences, the crucial importance of the definition of roles and tasks of different stakeholders involved in the process are outlined in the following:
- The elaboration of an efficient, adequate legal and regularity framework and the provision of budget lines are basic tasks of the central and regional governments, respectively
- Since sanitation programmes are far more than just hardware dissemination, the definition of the procedure for the institutionalisation process within public bodies (horizontal and vertical level) must be defined within the regularity framework. Special emphasis should be given to the definition of responsibilities and co-operation between different ministries and departments (public works, environment, health, etc); as well as how they are broken down at the national and local level
- Regional and local governments should be aware of the important role that sanitation programmes play within regional integral development. Sanitation goals and corresponding timelines should be established; these should comply with national legislation, norms and standards. Regional and municipal levels should monitor and ensure efficient co-ordination between concerned public entities.
- In most cases, the provision of sanitation facilities is the responsibility of local government, which must carry out the following tasks during implementation: awareness building within communities, decision-making in close collaboration with concerned communities, developing implementation schemes, budget allocation, monitoring implementation, setting up sludge-treatment systems and ensuring sustainability of the programmes
- Sanitation schemes must be developed in close co-operation with the communities. Since hygiene starts with the awareness and sanitation practices of each individual, sanitation programmes usually fail without the active involvement of the households. Community involvement is essential for ensuring regular use, continuous maintenance and financing of the sanitation facilities

Towards comprehensive wastewater and sanitation strategies

- Private-sector companies must not only deliver good-quality hardware, but also ensure long-term operation and maintenance as service providers. Public-private partnership models can ensure large-scale implementation and operation of sanitation facilities
- In many countries, non-governmental organisations (NGOs) initiate and facilitate the development and implementation of sanitation programmes. They launch awareness raising-campaigns, facilitate decision-making within communities, establish communication between communities and local governments, and even work as implementing agencies or service providers. Their roles depend on the profile and institutional competencies of each respective organisation, as well as the local conditions of the project area

Picture 2_15: Multi-stakeholder involvement is vital to successful sanitation programmes

Picture 2_16: "Demand-responsive approaches" have been developed in order to ensure the efficiency and sustainability of sanitation programmes

The concepts of "multi-stakeholder involvement" and "result-driven programme portfolio" correlate with a new perception of basic-needs infrastructural development:

- The "demand-responsive approach" should be a main feature of any sanitation programme; the users are perceived as "clients", who express a need and create a demand for sanitation services. Since public entities and other stakeholders respond to the demand of the communities, the approach is referred to as "Community-Based Sanitation"
- The active involvement of "users", "clients" or communities is crucial for the sustainability of the programme. "Willingness to pay" is not only a strong indicator that the community is actually interested in the programme, but also the basis for professional, long-term operation and maintenance of the sanitation system[19]

The demand responsive approach and the main principles of community-based sanitation are discussed in chapters 5 and 6 of this handbook

2.6 The increasing demand for efficient and reliable decentralised wastewater-treatment solutions

Sanitation programmes should be an integral part of comprehensive wastewater strategies and vice-versa. Connecting sanitation facilities to sewerage or to septic tanks alone, however, does not ensure the adequate treatment of domestic wastewater. In order to meet legal effluent standards, solutions for secondary and tertiary treatment must be found.

In recent years, improved legislation and growing public awareness have led to a rapidly growing demand for suitable wastewater solutions. Water quality and discharge standards are defined on the basis of legislation, such as the Philippines "Clean Water Act" (2002), the Vietnamese "Law on Water Resources" (1999), or the "Water Act" in India. These standards are subject to law enforcement, court cases and public debate.

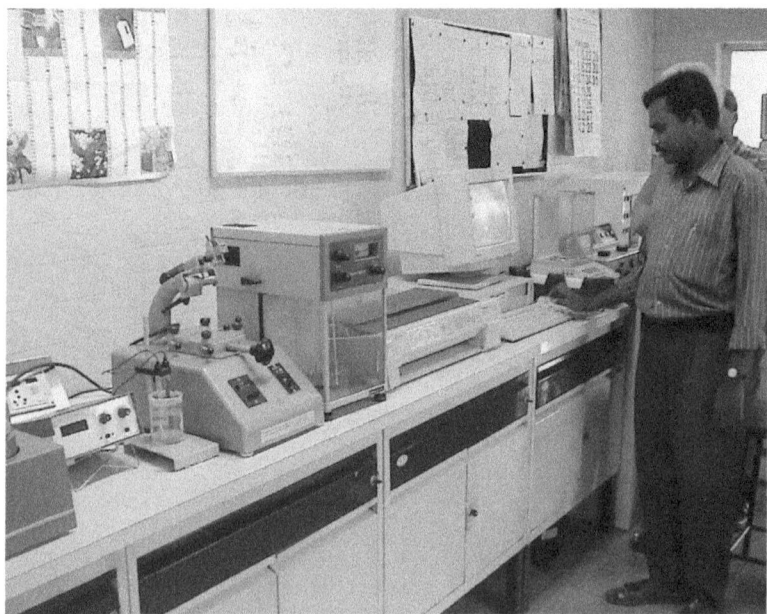

Picture 2_17
Improved legislation and law enforcement are the main driving forces behind the rapidly growing demand for new wastewater treatment solutions

DEWATS – Sustainable treatment of wastewater at the local level

Private and public entities are faced with the following situations:
- national and regional development plans require the wastewater connection of peri-, semi-urban and rural settlements to treatment facilities, which meet discharge standards
- new housing and real estate developments do not get clearance without approved wastewater-treatment systems
- schools, hospitals, hotels and public facilities face public pressure, due to surface-water pollution
- small and medium enterprises unable to treat wastewaters adequately are closed down by public authorities

Only a few of the households – well as public and private entities, that require wastewater treatment can be serviced by conventional sewage and wastewater-treatment systems. The rapidly growing demand can only be met with the assistance of other technical solutions, which should ideally fulfil the following criteria:
- suitable for very diverse local conditions and versatile in application
- provide reliable and efficient treatment of domestic and process wastewater
- require only short planning and implementation phases
- moderate investment costs
- limited requirements for operation and maintenance

It is evident that decentralised wastewater solutions, which fulfil these criteria, have to become an integral part of comprehensive wastewater strategies, complementing other approaches.

3.1 DEWATS – a modular system approach to ensure efficient wastewater-treatment performance

"Decentralised Wastewater Treatment Systems" (DEWATS) were developed by an international network of organisations and experts. In this handbook, the term DEWATS may be applied in singular or plural form, referring to a single specific system, to the modular systems approach or the whole range of systems, as the case may be. The approach incorporates lessons learned from the limitations of conventional centralised and decentralised wastewater-treatment systems, thereby assisting to meet the rapidly growing demand for on-site-wastewater solutions. DEWATS are characterised by the following features:

- DEWATS encompass an approach, not just a technical hardware package, i.e. besides technical and engineering aspects, the specific local economic and social situation is taken into consideration
- DEWATS provide treatment for wastewater flows with close COD/BOD ratios from $1m^3$ to $1000m^3$ per day and unit
- DEWATS can treat wastewaters from domestic or industrial sources. They can provide primary, secondary and tertiary treatment for wastewaters from sanitation facilities, housing colonies, public entities like hospitals, or from businesses, especially those involved in food production and processing.
- DEWATS can be an integral part of comprehensive wastewater strategies. The systems should be perceived as being complementary to other centralised and decentralised wastewater-treatment options
- DEWATS can provide a renewable energy source. Depending on the technical layout, biogas supplies energy for cooking, lighting or power generation
- DEWATS are based on a set of design and layout principles. Reliability, longevity, tolerance towards inflow fluctuation, cost efficiency and, most importantly, low control and maintenance requirements

DEWATS – Sustainable treatment of wastewater at the local level

- DEWATS usually function without technical energy inputs. Independence from outside energy sources and sophisticated technical equipment provides more reliable operation and, thereby, fewer fluctuations in effluent quality. Pumping may be necessary for water lifting
- DEWATS are based on a modular, technical configuration concept. Appropriate combinations of treatment modules can be selected, depending on the required treatment efficiency, costs, land availability, etc.
- DEWATS units are quality products. Though they can be constructed form locally available materials and can be implemented by the local workforce, high quality standards in planning and construction have to be met. For sound DEWATS design a good comprehension of the process of wastewater-treatment is essential
- DEWATS require few operation and maintenance skills. While most operational tasks can be carried out by the users, some maintenance services might require a local service provider. In some cases, both operation and maintenance can be delivered by a service provider
- DEWATS can reduce pollution load to fit legal requirements. Like all other wastewater-treatment systems, generated solid waste (sludge) must be handled, treated and disposed of in accordance with hygiene and environmental standards
- DEWATS consider the socio-economic enviroment of a given location. Neglecting these conditions will result in the failure of the technology

3.2 DEWATS – a brief insight into technical configuration

Typical DEWATS combine the following technical treatment steps in a modular manner:
- primary treatment – in sedimentation ponds, settlers, septic tanks or bio-digester
- secondary treatment – in anaerobic baffled reactors, anaerobic filters or anaerobic and facultative pond systems
- secondary aerobic/facultative treatment – in horizontal gravel filters
- post-treatment – in aerobic polishing ponds

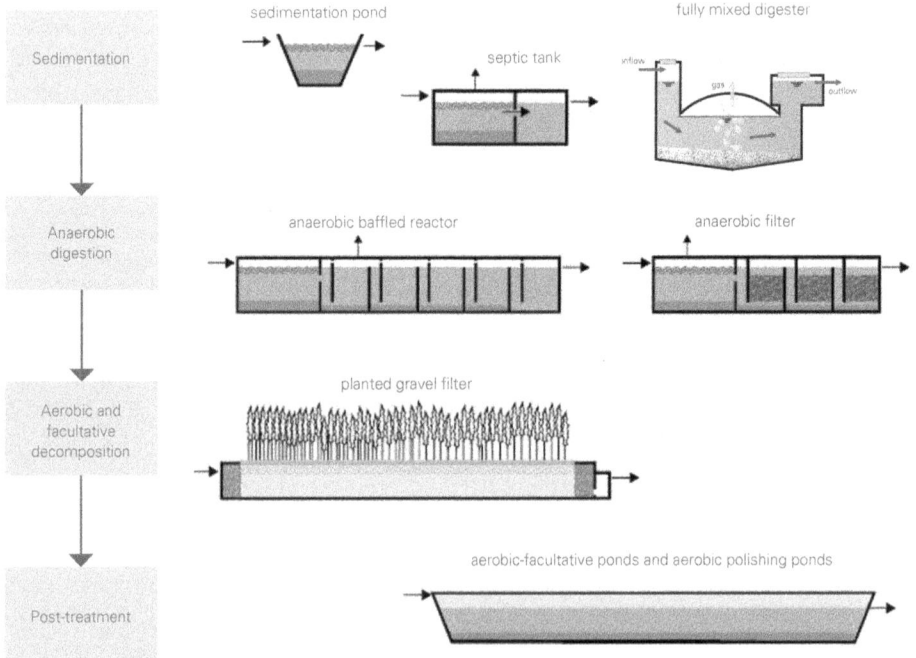

Picture 3_1: DEWATS configuration scheme

DEWATS – Sustainable treatment of wastewater at the local level

The selection of appropriate technical configuration depends on the:
- volume of wastewater
- quality of wastewater
- local temperature
- underground conditions
- land availability
- costs
- legal effluent requirements
- cultural acceptance and social conditions
- final handling of the effluent (discharge or reuse)

DEWATS rely on the same treatment processes as conventional treatment systems:

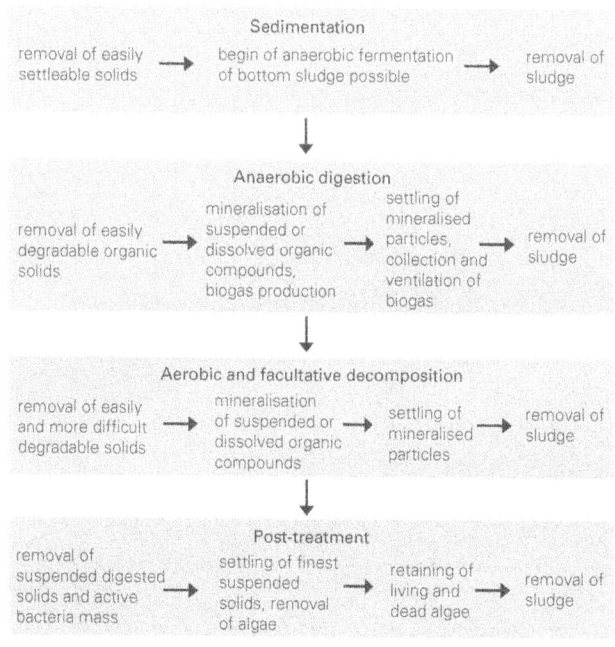

Picture 3_2:
Typical succession of treatment processes within DEWATS

3.3 DEWATS – good practice examples/applications

In recent years, DEWATS have been implemented at many different locations by various institutions. Gathered experience shows that each location demands its own approach. Below, a number of "good practice examples/applications" of DEWATS are presented. These are not meant to be exhaustive; they highlight different aspects of DEWATS implementation.

3.3.1 DEWATS/CBS – Community-Based Sanitation Programme in Alam Jaya, Tangerang, Java, Indonesia

Alam Jaya is a slum in the middle of an industrial area in Jakarta. Most residents work in the nearby factories. Due to a high migration rate, social structures are weak. The level of infrastructure development is low. Housing is poor with insufficient water supply.

Sanitation facilities in the settlement are totally insufficient in terms of quality and quantity. Wastewater is discharged into the environment without any treatment, posing a permanent threat to human health.

Picture 3_3: Housing in Alam Jaya

DEWATS – Sustainable treatment of wastewater at the local level

Bina Ekonomi Sosial Terpadu (BEST - Institute for Integrated and Social Development), a Tangerang-based non-profit organisation, has been promoting "Community Sanitation Centres" (CSC) since 1999. The centres provide basic sanitation facilities, such as toilets, bathrooms, a laundry area and "water points". The total wastewater flow is treated in a DEWATS. Until 2008, 33 Community Sanitation Centres have been implemented in the Tangerang and Surabaya areas, serving 14,800 users and treating 1,197m³ of wastewater per day.

Picture 3_4:
Typical sanitation facilities in Alam Jaya

Picture 3_5:
New Community Sanitation Centre in Alam Jaya

An intensive discussions process within the community preceded the decision to build a Community Sanitation Centre:
- the residents' desire for on-site toilets could not be met, due to the small size of the houses and plots
- the residents already use public toilets
- there was great interest in a reliable "water supply point"
- residents expressed their willingness to pay for water-supply and sanitation services

Picture 3_6:
Toilet at the CSC in Alam Jaya

The wastewater of the residents of the Alam Jaya quarter RT 02 RW 06 (65 households with 325 people) has the following parameters:

Source of water	domestic
Volume	37.5m³/day
Daily peak-flow hours:	16h
COD, influent:	743mg/l
BOD, influent	391mg/l
HRT in baffled tank	30h
Minimal digester temperature	30°C
Specific organic load (BOD_5):	0.34kg/(m³ × d)
Number of up-flow chambers	6 chambers
Volume of baffled reactor	49.39m³
COD, effluent	137mg/l
BOD_5, effluent	62mg/l

Table 2:
Data of Alam Jaya plant. in 2003, the construction cost were 167 Mio. IDR or 20,000 US$. Operation cost = 444,000 IDR or 55 US$/month. Users pay per use

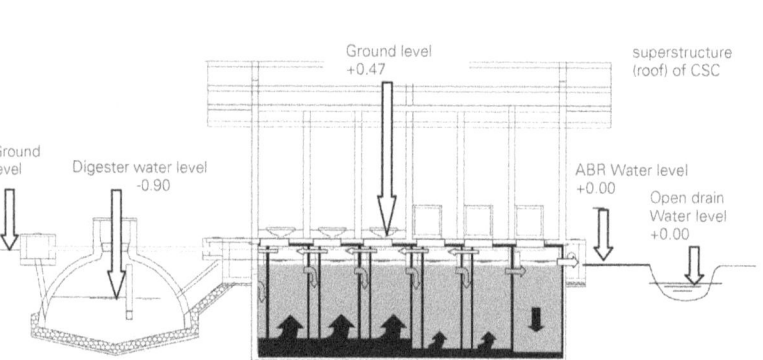

Flow separation of black and grey water:
- black water from toilets is treated in the bio-digester
- overflow and grey water from bathrooms is treated in the ABR

Picture 3_7:
Section of Community Sanitation Centre (CSC) in Alam Jaya with toilets and bathrooms

3 DEWATS – Sustainable treatment of wastewater at the local level

3.3.2 DEWATS/CBS – Community-Based Sanitation Programme in Ullalu Upanagara, Bangalore, India

Ullalu Upanagara is a peri-urban slum, located south-west of Bangalore, with 3,569 households and 17,325 people of different ethnic groups. The socio-economic situation of the residents is critical: inadequate basic amenities, high unemployment, low literacy. Women in particular face social hardship within their families and the community.

The weak socio-economic conditions are reflected in the infrastructure development. Access to reliable drinking-water supply, to proper housing and to clean sanitation is virtually non-existent. Only 21% of the households have their own toilet. The residents defecate openly – hindered by recent fencing.

Grama Swaraj Samithi (GSS), a local NGO, has been working in Ullalu Upanagara in the field of preventive health care since the 1990s. Since 2001, GSS has been promoting Community-Based Sanitation within the community. In close collaboration with the residents and local authorities, the construction of two sanitation centres was decided on. The implementation process was carried out as a pilot-programme, to test the application of participatory, administrative and technical instruments of the Community-Based Sanitation programme for the area.

Picture 3_8: Infrastructure is poor in Ullalu Upanagara

The participatory planning process resulted in the following layout of the overall complex:
- 2 separate sections – one for women, one for men
- 11 toilets and 1 bathing unit per section
- 12 laundry facilities – 8 for women, 4 for men
- fresh-water consumption:
  – 11.5m^3 per day
  – water connection and supply assured by Zilla Panchayat
  - use of rainwater harvesting tank during the rainy season
- source and quantity of effluent:
  – toilet and bathing wastewater: 7.5m^3 per day
  – laundry wastewater: 4.0m^3 per day
- low maintenance:
  – no piped water in toilets and bathing units
  – minimum electrical devices
- security
  – female and male sections visibly separated
  – entrance area for control and collection of service charges

Picture 3_9:
Community Sanitation Centre under constructio

DEWATS – Sustainable treatment of wastewater at the local level

Construction costs of the sanitation centre including DEWATS in 2004:
- sanitation unit: INR 550,000 (US$ 11,875)
- DEWATS-unit: INR 1,000,000 (US$ 21,590)
- bore well and electrical power connection: INR 275,000 (US$ 5,940)
- land value: INR 1,280,000 (US$ 27,640)
- total cost per complex: INR 3,105,000 (US$ 67,045)

 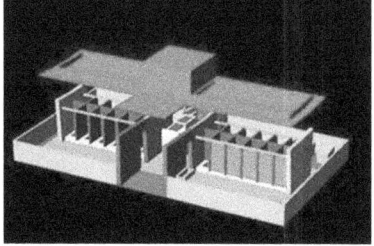

ture 3_10:
Newly inaugurated sanitation centre. Initially designed for 750 users per day, today over 1,000 visitors due to a nearby bus station.

ture 3_11:
Computer drawing of a sanitation unit

ture 3_12:
The Community-Based Sanitation programme was closely planned in collaboration with the future "users"

3.3.3 DEWATS at public institutions – Sino-German College of Technology, Shanghai, China

The Fenxian campus of the Sino-German College of Technology at East China University of Science and Technology is located an hour's drive from Shanghai. It is an engineering college and its campus was planned for 6,500 teachers and students (no accomodation).

The challenge for the school's authorities was to find a reliable and efficient solution for treating their wastewater in accordance with the Environmental Standard GB/T 18921-2002 (2nd stage). Tight budget constraints for initial investment and operation restricted the possible wastewater-treatment options.

The campus wastewater consists of toilet effluent from the teaching buildings, as well as polluted water from machinery-maintenance processes. The DEWATS technical configuration had to consider therefore oil, NH_3-N, grease and swarf, besides the normal parameters of COD and BOD_5.

Picture 3_13:
View of the college campus

DEWATS – Sustainable treatment of wastewater at the local level

The chosen DEWATS consists of a module for grease separation and sedimentation, a three-step anaerobic digester with filter, an underground sand filter (biofiltration) and an irrigation tank. Operation started in September 2004. The effluent is used to irrigate compound gardens, while biogas is used to light campus street lamps and water heating. The project costs were calculated at 960,000 RMB (US$ 115,942).

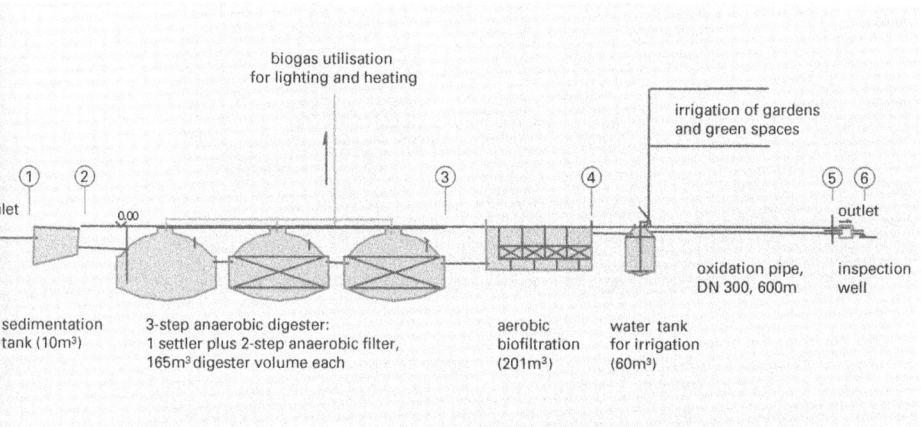

Figure 3_14:
Schematic drawing of the DEWATS solution at the Sino-German College of Technology in Shanghai, Fengxian District

Picture 3_15:
View of campus buildings and biogas street light

Picture 3_16:
DEWATS under construction

The effluent of the plant shows that the required discharge standards are met:

Sample Point	Inflow	Sedimentation tank	3-step anaerobic digester	Aerobic sand filter	Aerobic oxidation pipe	Inspection well	Required legal standard
	1	2	3	4	5	6	
Daily wastewater flow [m³]	146.25	146.25	146.25	146.25	146.25	146.25	
Capacity [m³]		10	495	195	5	1	
HRT [h]			81	32	0.8		
COD_{cr} [mg/l] (removal rate)	800	720 (10%)	108 (85%)	91.8 (15%)	87.21 (5%)	87	100
BOD_5 [mg/l] (removal rate)	400	360 (10%)	39.6 (89%)	31.68 (20%)	28.5 (10%)	28.5	30
SS [mg/l] (removal rate)	200	180 (10%)	90 (50%)	45 (50%)		45	150
NH_3-N [mg/l] (removal rate)	80		40 (50%)	16 (60%)	14.4 (10%)	14.4	15
Oil [mg/l] (removal rate)	20	10 (50%)				10	15

Table 3:
Water-treatment data (analysis by local environmental protection bureau)

3.3.4 DEWATS at public institutions – Aravind Eye Hospital in Thavalakuppam, Pondicherry, India

The Aravind Eye Hospital in Thavalakuppam belongs to the Tamil Nadu-based Aravind Eye Care System. The philosophy of the Aravind System is to provide services to the rich and poor alike, while achieving financial self-sustainability. This is achieved through high-quality, large-volume care and efficient management.

The hospital in Thavalakuppam has the capacity to treat 750 in patients (600 free admissions and 150 paid) and an additional 900 out patients. 300 paramedical staff are housed in 26 residential quarters.

Due to the water scarcity in the region, the hospital management expressed strong interest in a wastewater-treatment solution, that permits the reuse of treated water.

The chosen DEWATS solution was designed to treat approximately 307m³/d of domestic wastewater from toilets, bathrooms and kitchens. Water reuse (due to high water scarcity) and efficient land use had the highest priority in treatment-process selection.

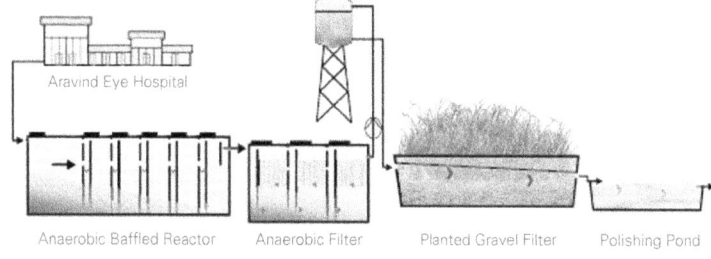

Picture 3_17:
Schematic drawing of the DEWATS at Aravind Eye Hospital

The effluent of the DEWATS-plant irrigates a garden with 300 trees planted in avenues, 250 coconut trees, 50 mango trees and 4,200m² of lawns, covered with Korean grass and flowering plants. In 2004, the hospital was honoured with the Pondicherry Government's award for the best garden. Construction started mid 2002, start of operation was February 2003. Construction cost are 10 Mio INR (200,000 US$).

Picture 3_18:
Polishing pond of Aravind Eye Hospital's DEWAT Through reuse of treated wastewater, Aravind Eye Hospital saves annually 100,000m of freshwater.

DEWATS – Sustainable treatment of wastewater at the local level

Picture 3_19:
Horizontal filter with *canas indica*, *reed juncus* and papyrus plants

Picture 3_20:
Baffled reactors are used as a parking lot

3.3.5 DEWATS/SME-Cluster approach – Kelempok Mekarsari Jaya
small-scale industry cluster, Denpasar, Bali, Indonesia

Mekarsari Jaya is a small-scale industry cluster in Pucuksari Selatan, Banjar Batur, Denpasar. It consists of 54 entrepreneurs, engaged in tofu production and chicken slaughtering. At the same time, Mekarsi Jaya is a settlement area for migrants from other parts of Indonesia. Due to the poor infrastructural conditions, the area is considered a "slum" by the local residents.

Wastewater from domestic and industrial sources is generally discharged to nearby "dead water" channels without any treatment. But recently enforced environmental regulations, mean that enterprises are forced to treat their wastewaters before discharge.

The project was planned and implemented by BaliFokus, a Denpasar-based NGO. Due to the settlement structure and topographical condition of the area, the implementation of a central treatment unit for Mekarsari Jaya faced major technical obstacles. In order to meet the legal requirements of the authorities, it was decided to implement two DEWATS in the area. While one system in Northern Pucuksari serves 11 tofu-processing units and 5 chicken-slaughter houses, a second system in Southern Pucuksari Selatun serves 7 processing plants.

Picture 3_21:
Tofu processing causes high water pollution in Mekarsari Jaya

Picture 3_22:
Domestic and industrial wastewa is discharged to channels without treatment

Wastewater analysis shows high loading of the wastewater:
- Northern unit – 50m³/d wastewater influent with a BOD of 7,000mg/l and COD of 11,000mg/l.
- Southern unit – 20m³/d wastewater influent with a BOD of 5,000mg/l and COD of 8,000mg/l.

Topography and settlement structure (densely populated) were the decisive factors for technical plant layout. A bio-digester, followed by an anaerobic filter, were found most suitable to treat the highly loaded wastewater.

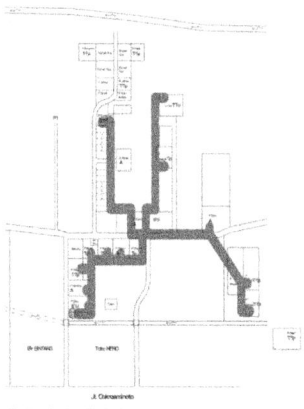

Figure 3_23: DEWATS treats wastewater from several industrial units (sewerage system= blue lines)

The following data characterises the DEWATS solution of the northern unit:

Source of water	domestic
Volume	50m³/d
Daily peak-flow hours:	12h
COD, influent:	11,000mg/l
BOD_5, influent	7,000mg/l
HRT in anaerobic filter	17.5h
Minimum digester temperature	30°C
Number of up-flow filter chambers	3 chambers
Volume of baffled reactor	36.45m³
COD, effluent	335mg/l
BOD_5, effluent	191mg/l

Table 3a:
Characteristics of DEWATS solution for small scale food processing industry and settlement at Mekarsari Jaya (northern unit).
Start of construction 09'2003
Start of operation 04'2004
Construction cost: 112 Mio. IDR (US$13,500)

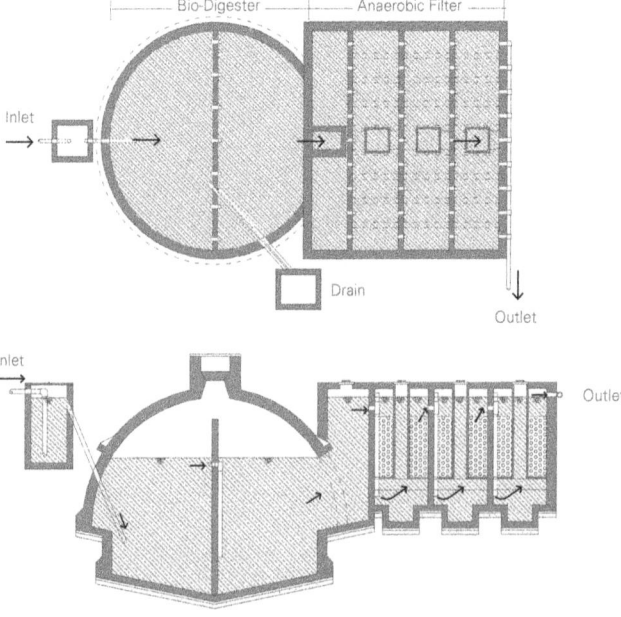

Picture 3_24:
Technical layout and view of the DEWATS unit at Mekarsari Jaya (northern unit)

3.3.6 DEWATS/SME –
Alternative Food Process Private Ltd. Bangalore, Karnataka, India

The food-processing unit is located in the suburbs of Bangalore city. The company operates a gherkin-processing plant, where selected gherkins are washed, prepared, pickled and stored over a period of 12 days before export.

The company caters semi-finished products to leading brands. High quality production and adherence to delivery standards of international markets are the top priority. The company employs around 100 people and handles 8 to 10 tonnes of gherkins per day.

Picture 3_25 and 3_26: The products of Alternative Food Process Private Ltd. meet high delivery standards for national and international markets

The treatment of 29.1m³/d organic wastewater (COD 800 / BOD 400mg/l) is required. Due to water shortages in the area, water reuse is desirable.

To find the best treatment solution, a comprehensive analysis of the different wastewater streams was undertaken. By handling certain wastewater streams separately, the right treatment solutions could be applied to each situation:

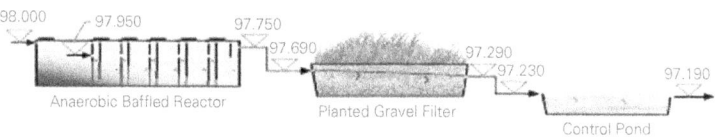

Picture 3_27: Schematic drawing of the DEWATS at Alternative Food Process Ltd.

Picture 3_28:
 The anaerobic filter under construction

Picture 3_29:
 View of polishing pond with shallow sections for better UV-disinfection and multi-levels for better aeration

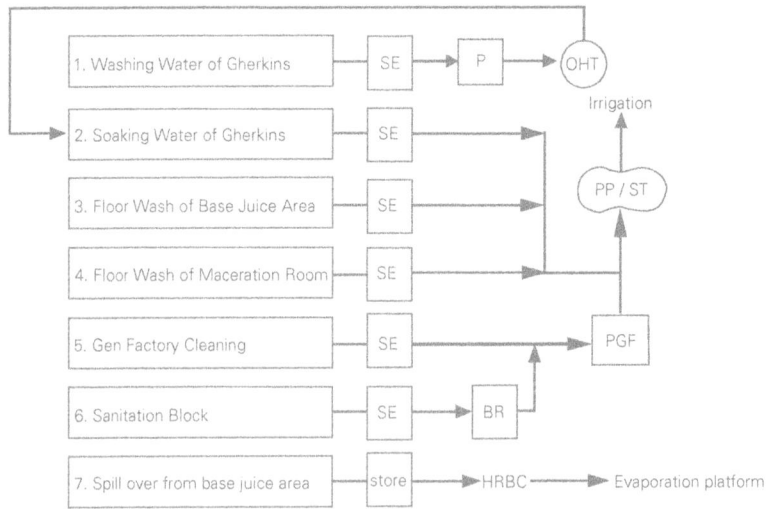

Picture 3_30:
 System layout (SE: Settler, BR: Baffle Reactor, PGF: Planted Gravel Filter, ST: Storage Tank, PP: Polishing Pond, HRBC: High-rate Brine Condenser (evaporation), OH Overhead tank, P: Pump)

3.3.7 Infrastructural development in rural China – Longtan Village, Danleng County, Szechuan Province, China

The Chinese government aims to improve rural livelihood by promoting the enhancement of rural infrastructure through different public programmes. Road construction, housing, electricity provision, biogas utilisation, water supply and wastewater schemes – as well as solid-waste management – are part of multiple village modernisation programmes.

Longtan Village has a population of 965 people living in 262 households. Agricultural production on approximately 56.7 ha of land is the main income source for the residents. Traditionally, paddy and oil seeds were cultivated. However, economic reforms have brought significant changes to Longtan: the village has begun market production of oranges, grapes and oil seeds, while raising 1,250 pigs in 2005. In this year, a household's average annual income was about 3,113 RMB per person (US$ 420).

Public authorities in rural China have the challenge of meeting legal wastewater discharge standards. New air-quality standards have also been issued, demanding a different treatment of rice-harvest residues, which were traditionally burned. As a result, decentralised wastewater-treatment systems are promoted. A combination of anaerobic and of aerobic-treatment units is applied to treat animal dung, human faeces and residues from agricultural production. Biogas provides a renewable-energy source, while slurry can be used in organic farming.

Picture 3_31: DEWATS treats human faeces and agricultural residues

Picture 3_32: DEWATS-generated biogas is used for multiple purposes, such as water heating

The village's development plan stipulates that 120 households should be connected to biogas units, each with a volume of 10m³. Rice residues are processed in a chaff cutter before being emptied into the digesters. Bio-digesters with a volume of 3.5m³ are mandatory for households without paddy production. Where possible, homes are connected to one of two DEWATS plants in the village. The treated wastewater is discharged into the open drainage system, which crosses the village.

Picture 3_33:
Infrastructural development programmes aim to modernise Chinese villages

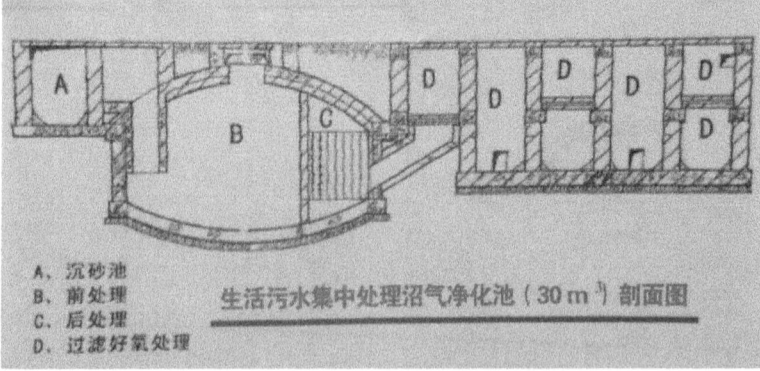

Picture 3_34
DEWATS – settler, bio-digester, anaerobic baffled reactor and horizontal filters (not shown)

DEWATS – Sustainable treatment of wastewater at the local level

3.3.8 DEWATS in integrated municipal planning – Wenzhou University, Zheijang Province, China

Since the 1980s, the government of Zhejiang Province has been promoting DEWATS, particularly in urban areas, which are not connected to centralised systems. Today, many of the province's sources of domestic wastewater, such as public toilets, apartment buildings, schools, hospitals and universities are served by these treatment systems. Apart from domestic applications, decentralised wastewater-treatment solutions are applied at small- and medium-scale enterprises, like slaughterhouses, food processing and animal-husbandry units.

The Wenzhou New Energy & Environmental Design Institute (WNEEDI), an Institute of the Rural Energy Office Wenzhou, is active in the dissemination of innovative renewable-energy and ecological wastewater-treatment projects (biogas plants, DEWATS, solar thermic systems, hydro rams) within the city and Wenzhou County. WNEEDI started by promoting biogas plants 50 years ago and has slowly shifted its main activities to wastewater treatment in urban areas.

Figure 3_35:
The central administration building of the University of Wenzhou

Figure 3_36:
Arial view of the University of Wenzhou campus

Within this context, WNEEDI was responsible for the planning and implementation of an integrated wastewater concept for Whenzhou University, the first university run jointly by the government and business. In 2005, the university had approximately 10,000 students.

The DEWATS implemented at the University campus are viewed as the ideal long-term solution. The treatment facilities will grow incrementally, in line with the addition of new buildings and the overall growth of the campus.

Today, the university uses multiple DEWATS, with a total reactor volume of about 90,000m³. Nearly all buildings, including the dormitories, have their own primary treatment unit, which connects to shared, secondary treatment units. Units of approximately 20 different treatment volumes, ranging from 40 to 800m³, have been implemented.

All systems consist of pre-treatment in fixed dome biogas modules. Two to four digesters are usually connected in series. After anaerobic treatment, the wastewater is aerobically treated by flowing over cascades. Final treatment is provided by two to four horizontal-flow sand filters in series.

Implementation is carried out by contractors, specialised in decentralised wastewater treatment. To ensure gas-tight construction of biogas domes, certification of the building contractors is required. The local Rural Energy Offices are responsible for certification; Wenzhou County has eight certified contractors.

Picture 3_37:
The project team tests the treatme performance.

Picture 3_38:
Construction of anaerobic filter (in front)

Mainstreaming DEWATS – strategic planning and implementation of sustainable infrastructure

Nowadays public authorities are challenged to provide sanitation and wastewater-treatment services on a large scale. Mainstreaming decentralised wastewater-treatment solutions is one of the key elements for sustainable infrastructure development.

4.1 Strategic planning of sanitation programmes

Comprehensive wastewater strategies may consider different options for the treatment and discharge of wastewater:
- treatment in a centralised plant, which is connected to a combined or separate sewer system
- treatment in several medium-sized treatment plants, which are connected to a combined or separate sewer system
- primary and secondary treatment in decentralised plants, which are connected to a sewer line, leading to a common plant for final treatment
- completely decentralised treatment with final discharge, reuse, or connection to communal sewerage
- controlled discharge without treatment (ground percolation, surface-water dilution)

The final decision, on which treatment option is most suitable for a given water pollution problem, should be based on a number of different considerations, which are discussed in greater depth later in this book. Different options may be considered for residential areas:
- Simplified community-sewerage systems with household-based sanitation systems are preferred in areas where the residents have sufficient financial resources and households have sufficient space. On average, 20 to 100 families are connected to one system. The system consists of toilets and bathrooms within each household. The wastewater is directed to a DEWATS by shallow, narrow sewer lines.
- Shared septic tanks present a simpler version of the household-based sanitation system with off-site treatment. A smaller cluster of about 10 to 50 households is connected to a community septic tank. The system treats toilet and bathroom effluent from each household. Wastewater is channeld to the septic tank by shallow small-diameter sewer lines. The wastewater cannot be discharged directly to the aquatic environment, due to the low effluent quality of the septic tank. The system is, usually only applied, therefore where soil conditions allow the direct infiltration of the effluent without any harm to the groundwater.

- Community Sanitation Centres (CSCs) are appropriate in areas where financial resources are very limited and most residents live in rented rooms or huts, leaving no space for in-house sanitation. The centre is established at a central location within the settlement and offers different services as requested by the community. Services can include water points, toilets, bathrooms and laundry areas. Each CSC is connected to a DEWATS, usually located underground below the Centre. CSCs are usually guarded and operated by paid staff.

The experience gathered in multiple efforts to create efficient and cost-effective sanitation and wastewater-treatment strategies clearly shows that, without comprehensive legal frameworks and efficient law enforcement, without institutional capacities within public and private services, without relevant financial resources, and without awareness at the household or enterprise level, the hoped-for health and environmental standards cannot be achieved.

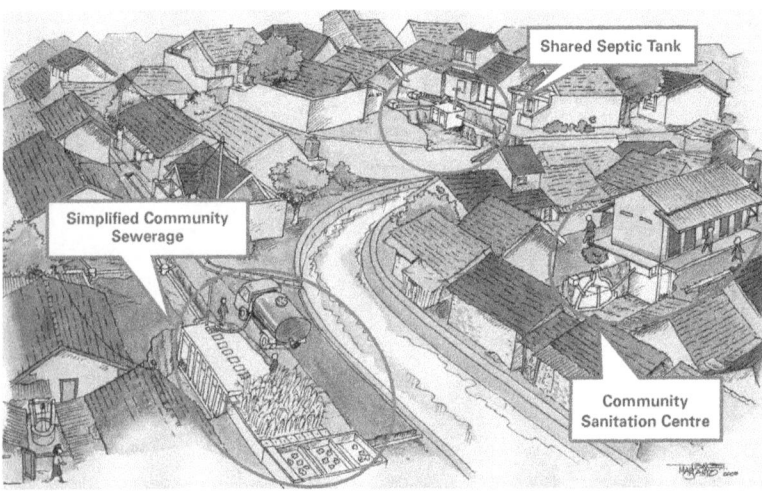

Picture 4_1: Different treatment options within a CBS programme

Mainstreaming DEWATS – strategic planning and implementation of sustainable infrastructure

In many countries, new legal regulations have favoured a rapid increase in the demand for decentralised wastewater-treatment systems. For many public and private entities DEWATS poses the only solution for complying with legal requirements within the time constraints. The situation raises the question: How can these technical options be integrated effectively into regional and municipal planning processes, in order to reach an economy of scale?

Since the goal of public authorities is not to promote specific technical solutions, but rather to achieve political and administrative targets, the following questions must be considered by all key decision-makers:
- Under which conditions should DEWATS be preferred over other technical solutions?
- What are the advantages of DEWATS over other wastewater-treatment options?
- How can a legal and institutional framework be created, which facilitates comprehensive sanitation and efficient wastewater-treatment schemes?
- What are the core elements of such schemes?
- Who are the stakeholders, who should be involved in the process?
- What kind of approach is required to ensure efficient, cost-effective and sustainable implementation?
- How can the implementation of such schemes be initiated and maintained?

The government of Indonesia, for example, evaluated multiple efforts in the sanitation sector, as a basis for creating an implementation scheme for a nation-wide programme. It was concluded that the exclusive top-down approach must be replaced by a conceptual framework, which includes "demand-driven services", "multi-stakeholder involvement", and "multi-task planning" as guiding principles.

In order to overcome the poor long-term performance of many projects and initiatives, the government of Indonesia has decided that further guiding principles should play an integral part in any planning and implementation activities:
- sustainability of financing
- sustainability of technical know-how
- sustainability of environmental management
- sustainability of infrastructural management
- sustainability of social interaction

This conceptual framework reflects the international discussion about how sanitation and wastewater treatment services can have the optimal sustainable impact.

The following features present the underlying principles of an efficient and cost-effective programme:
- comprehensive legal regulations
- efficient law enforcement
- target-orientated local and municipal planning
- demand-responsive approach
- comprehensive assessment of local and community needs
- service orientation
- multi-stakeholder involvement
- appropriate choice of technology
- multi-task planning
- financial analysis and long-term financial planning
- sound planning and monitoring of the implementation process
- capacity building
- step-by-step implementation

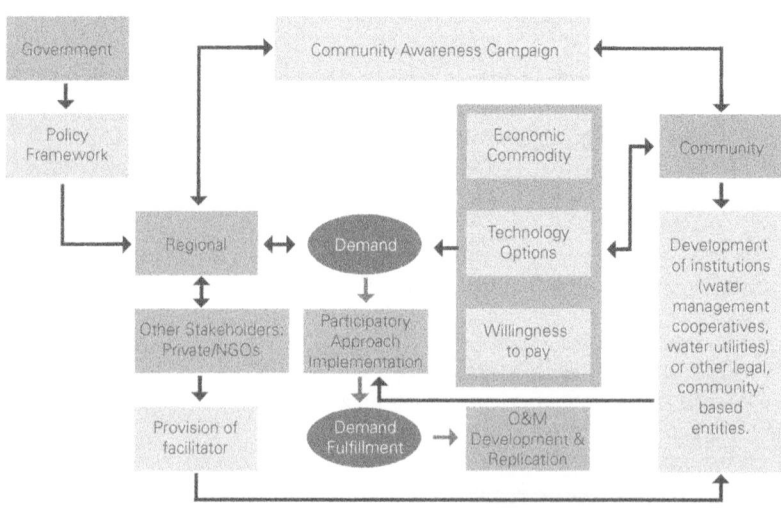

Picture 4_2: Conceptual framework of Indonesian National Sanitation Programme

4.2 Legal framework and efficient law enforcement

A comprehensive legal framework and its efficient enforcement at the local level are essential to the success of sanitation and wastewater-treatment strategies.

Wastewater-treatment schemes must meet the legal discharge standards, defined within the legislation of each country. Those standards, however, are rarely met in developing countries. The reasons for this are manifold.

In most countries, legal environmental and discharge standards are based on the most scientifically advanced treatment technologies available on the market. Discharge standards in developing countries often refer to those from industrialised countries, where sophisticated treatment technologies can be applied to treat the highly diluted municipal sewage. The different prerequisites in developing countries, including wastewater composition, economic and socio-economic conditions as well as financial and organisational restrictions, create large discrepancies between desired effluent standards and the actual services that can be provided. In some cases, standards thereby achieve adverse effects, as they are considered unrealistic and ignored.

examples	COD g/cap.*d	BOD_5 g/cap.*d	COD/ BOD_5	SS g/cap.*d	Flow l/cap.*d
India urban	76	40	1.90	230	180
USA urban	180	80	2.25	90	265
China pub. toilet	760	330	2.30	60	230
Germany urban	100	60	1.67	75	130
France rural	78	33	2.36	28	150
France urban	90	55	1.64	60	250

Table 4:
Some selected domestic wastewater-data.
Source: BORDA

At point-source effluent sources, like hospitals and small-scale industries, compliance with given discharge standards often proves too expensive. Thus, individual polluters frequently decide to either completely ignore the problem or to set up a fake treatment system to please the environmental authorities. In other cases, complicated technology is implemented, but often soon results in the described performance problems.

Indian National Discharge Standards					
		discharge into			
parameter	unit	inland surface water	public sewers	land for irrigation	marine coastal area
SS	mg/l	100	600	200	100
pH		5.5 to 9	5.5 to 9	5.5 to 9	5.5 to 9
temperature	°C	<+5			<+5
BOD_5	mg/l	30	350	100	100
COD	mg/l	250			250
oil and grease	mg/l	10	20	10	20
total res. chlorine	mg/l	1			21
NH_3-N	mg/l	50	50		50
N_n-as NH_3	mg/l	100			100
free ammonia as NH_3	mg/l	5			5
nitrate N	mg/l	10			20
diss. phosphates as P	mg/l	5			
sulphides as S	mg/l	2			5

Table 5:
Source: Central Pollution Control Board, Delhi

It is becoming increasingly apparent that a more realistic approach must be sought:

> "Undue haste in adopting standards, which are currently too high, can lead to the use of inappropriate technology in pursuit of unattainable or unaffordable objectives and, in doing so, produces an unsustainable system. There is a great danger in setting standards and then ignoring them. It is often better to set appropriate and affordable standards and to have a phased approach to improving the standards as and when affordable. In addition, such an approach permits the country the opportunity to develop its own standards and gives adequate time to implement a suitable regulatory framework and to develop the institutional capacity necessary for enforcement."[20]

[20] Johnson et all, Institutional Developments, Standards and R Quality, 1996

Recently, an increasing number of countries have launched initiatives to draft more realistic legal frameworks. Regulations cover a wide range of topics, including the practices of service providers, design standards, tariffs, discharge standards and contracts. These regulations, especially design and discharge standards, are carefully adapted to local conditions and no longer just copied from regulations applied in industrialised nations.

For example, in its Water Act the Government of the Republic of South Africa defines differentiated wastewater-treatment and disposal standards, according to wastewater type, quantity and the location of generation. While high standards are applied to areas of high risk, in terms of ecology and health, lower standards are defined for other locations, such as sparsely populated areas. This pragmatic approach widens the scope of applicable technological solutions, ensuring a more site-specific treatment-option selection and thereby increasing the positive impact on health and environment on a larger scale.[21]

Ref. Government Gazette No. 20526 8 October 1999.

Comprehensive law enforcement was and is one of the major challenges to the successful implementation of wastewater strategies. Due to weak institutional capacities, the adherence to regulations was and is seldom properly monitored by public bodies. In many countries, the relevant authorities are rarely prepared to carry out performance-orientated site monitoring. Public agents frequently request the implementation of sophisticated hardware, even in cases where a decrease of wastewater pollution might be more efficient and less expensive achieved by wastewater-prevention measures. There is a great necessity for institutional capacity building. On the other hand, the corruption in many countries must be overcome, if the legal framework is to be enforced effectively.

The enforcement of comprehensive legal standards can be perceived as a major driving force for improving the current sanitation situation with efficient and cost-effective wastewater solutions:
- existing small and medium-scale industries are urged to comply with discharge standards in short term
- new industrial sites, slaughterhouses and hospitals only receive clearance once reliable wastewater treatment is provided
- new housing colonies and residences are only approved if they ensure efficient treatment of the generated domestic wastewater
- municipalities and local governments are urged to protect surface and groundwater bodies from the intrusion of domestic and industrial wastewaters

A reliable legal framework must be backed up by an efficient policy framework and law-enforcement procedures. Institutional capacities must be created, and standardised law-enforcement procedures must be developed.

Awareness-building campaigns within the civil society can help to create leverage for law enforcement. In many countries, cases have been filed by individuals, neighbourhood groups and NGOs, forcing polluters to close down operations because they were not willing or not able to meet discharge quality standards.

It seems obvious that recent and future ecological developments will be reflected in the legal frameworks. The extensive use of natural resources requires more stringent regulations. Surely economic instruments on the macro level will influence the sanitation sector in the near future. The more fresh-water resources are perceived as a valuable and scarce public asset, the higher water will be priced. Pricing directly influences water consumption and the search for wastewater solutions, which are based on reuse or "closed loop" concepts.

4.3 Target-oriented local and municipal planning

4.3.1 Features of urban infrastructure development

A closer look at the socio-economic structure of a city can provide a first overview of relevant decision parameters for the final selection of appropriate technology.

The dynamic economic growth of most cities in the developing world caused deep social transformation within these societies. Rural areas and villages were rapidly integrated into spreading urban settlements. Agricultural land was converted into new residential and industrial areas. These trends can be still observed almost everywhere.

In most cases, this development lacks systematic planning of land use and adequate infrastructural development. By studying urban land-use patterns, one can gain insight into the social stratification and economic diversification of an area. While "wealthy neighbourhoods" are supplied with relevant infrastructure, informal settlements are left with only limited or no access to basic infrastructure. Even if a central sewage system cannot be extended everywhere, "formal" settlements are usually connected to septic tanks, while "informal" settlements have no treatment at all. Wastewater from industrial areas is commonly channelled directly to the closest surface waters.

Since informal settlements are a major driving force in most urbanisation processes, the following land-use pattern is quite common in the larger cities of developing countries.

Close to the city centre, a number of informal settlements exist. These are often found in so called "risk areas", such as dump sites, railway crossings, etc. The livelihood of the dwellers is usually dependent on activities in the informal sector or day labour.

Similar "peri-urban" settlements are found at the outskirts of urban areas. Dwellers of these settlements commonly generate income from day labour and small-scale commercial activities or business. If possible, self-subsistence farming is practised on nearby lands. Due to the unclear situation regarding land ownership – and the general negligence towards the urban poor, there is little public investment in infrastructure in these regions.

Since these areas are most responsible for urban growth, their importance to comprehensive urban-infrastructure development is obvious. Particular entities, such as small-scale industry clusters, schools and hospitals in semi- and peri-urban areas, face the greatest problems in meeting discharge standards.

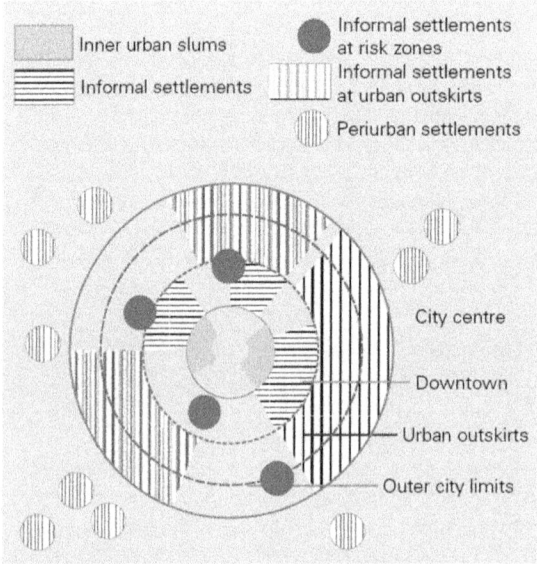

Picture 4_3:
Informal settleme
in greater urban
areas of developi
countries
Source: GTZ, 20C

4.3.2 Sanitation mapping as a tool for efficient urban-infrastructure development

In recent years Geographic Information Systems (GIS) have become an integral part of comprehensive urban-development strategies. GIS is a tool for visualising parameters, which are relevant for infrastructure development. Sanitation mapping permits the analysis of collected data, like the current situation of sanitation infrastructure, the impact of poor infrastructure on environment, and the driving forces, such as the socio-economic dynamics, of a given location. A database of the following parameters is beneficial for efficient sanitation mapping:
- topography
- natural water-drainage systems – rivers, streams, creeks
- land-use patterns – residential zones, industrial and agricultural areas
- existing city master-plan
- existing water-related infrastructure – sewers and water supply
- main water-pollution sources
- residential structure
- population density
- socio-economic situation of residents
- existing sanitation facilities
- community health conditions

GIS can be a powerful tool for poverty-alleviation programmes. Shelter Associates, an NGO based in Pune, India, implements housing programmes in poor areas. Shelter Associates applies GIS to generate a reliable database, which supports systematic programme approaches, as practised at the "Community Water and Sanitation Facility" at Sangli-Miraj-Kupwad Municipal Corporation (SMKMC).

SMKMC is located on the banks of the Krishna River in southern Maharashtra. In 2001, SMKMC had a population of 478,500. It covers about 118 km². Although the Municipal Corporation is only four years old, almost 15 per cent of its population live in slum settlements. The lack of access to basic infrastructure and civic amenities is a main feature of the area.

In order to get an overview of the existing sanitation situation, all the SMKMC settlements were mapped by plane table survey methods and each household was surveyed individually. The information was entered in GIS software and a detailed analysis of each slum pocket was compiled. The data generated gave a detailed picture of the existing water and sewage situation. Maps of the dilapidated water-supply network and sanitation facilities were developed.

It became apparent that the city has not undertaken any major improvements or extensions in the past 20 years. As a result, more than 11,500 households in Sangli were left without any basic sanitation access. Information gathered on a household level underlined the linkage between poor sanitation and weak socio-economic structures.

Picture 4_4:
Satellite photo o
SMKMC (Source
by google earth)

Mainstreaming DEWATS – strategic planning and implementation of sustainable infrastructure

Figure 4_5:
Existing sewage system in SMKMC

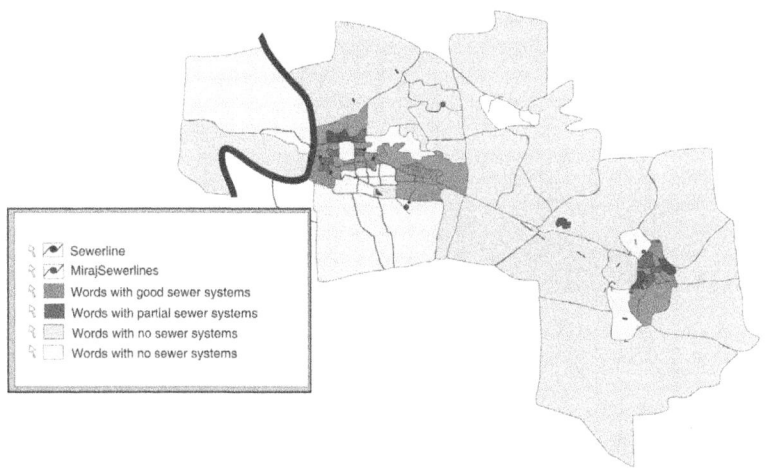

Figure 4_6:
Sanitation-relevant data on household level in SMKMC

Similar features can be observed in the City of Lusaka, Zambia. A GIS-generated sanitation map shows that there are two main sewage-disposal methods within the urban area:[22]
- centralised waterborne methods, which comprise a sewer network, sewage-pumping stations and sewage-treatment works and
- on-site sanitation methods, like septic tanks and soakaways, pit latrines, aqua privies and cesspools

Additional information, relevant to the future development of a wastewater strategy and the identification of suitable technological options, was compiled:
- only about 30% of the areas, which receive water supply from the Lusaka Water and Sewerage Company, are serviced by a sewer network
- the sewer network is divided into six catchment areas, each serviced by a sewage-treatment plant
- storm water and sewage waste are drained through separate systems.
- the sewage network operates mainly on gravity flow; few areas are served by pumping stations

[22] Urban Development Plans and Infrastructure Services for the City of Lusaka., Lusaka Water and Sewerage Company, 2005

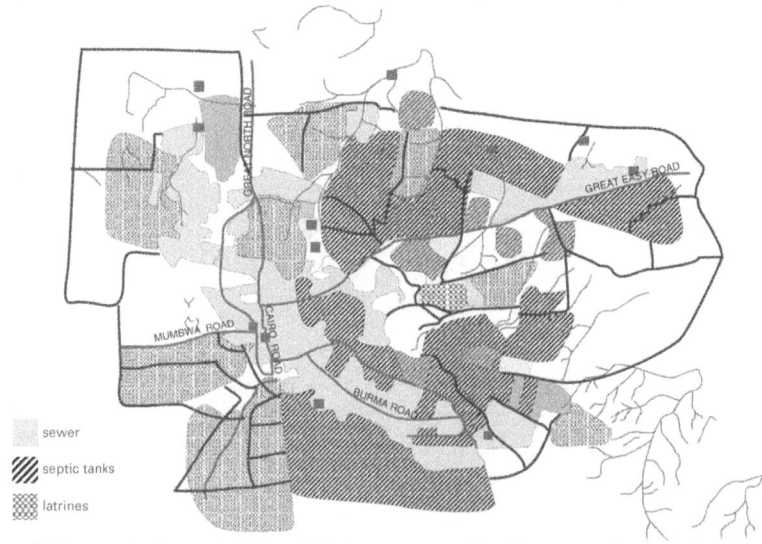

Picture 4_7: Lusaka Sanitation Map

Mainstreaming DEWATS – strategic planning and implementation of sustainable infrastructure

The Community-Based Sanitation programme in Ullalu, Upanagara, Bangalore, India, described in section 3.3.2 also applied sanitation mapping to find the most effective way to improve the sanitation situation of the large slum's dwellers. Besides the careful assessment of physical parameters, such as topography, land availability and existing infrastructure facilities, comprehensive household surveys were carried out. After detailed analysis, participatory methods for project planning were applied. The combination of physical and social data within the same maps showed the connections between the availability of sanitation facilities at the household level, the socio-economic situation of the dwellers and their preparedness to contribute to the overall improvement of sanitation infrastructure (willingness to pay). The insights-gained were key decision parameters for sanitation-centre site selection. Chosen sites provide both the required physical preconditions as well as a strong acceptance of the new utility by the users.

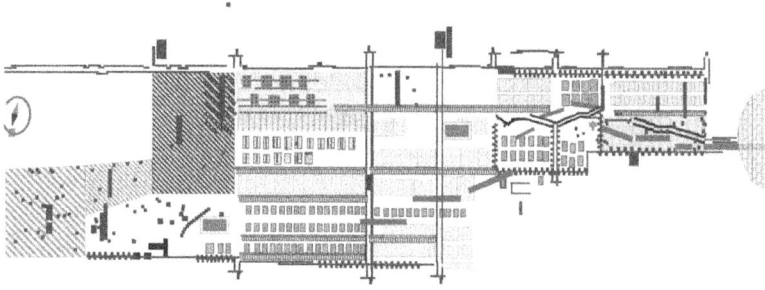

Figure 4_8:
Position of two sanitation complexes within Ullalu slum

For application in full-scale urban planning, sanitation mapping must combine a wide range of relevant parameters. Besides all the data mentioned above, the overall dynamics of current developments and the available resources within the sanitation sector should be included. The tool can then be used to assess whether decentralised wastewater-treatment solutions are appropriate for a given location.

The following locations are the most favourable types for the implementation of DEWATS:
- locations far away from central sewerage and wastewater-treatment systems, or where a connection to such a system is unlikely due to financial reasons
- locations suffering from water scarcity

- locations which are difficult to attach to central sewage systems, due to the topographical profile of the area (hilly areas, ravines, etc.)
- locations with polluters, such as schools, hospitals, slums, new housing colonies, and small and medium industries, needing immediate and intermediate wastewater-treatment solutions[23]

23 Further planning details will be discussed in chapter 5 and 6

A sanitation map – containing all relevant data and parameters – should help identify those areas of a city, that are most suitable for centralised and/or decentralised wastewater-treatment approaches.

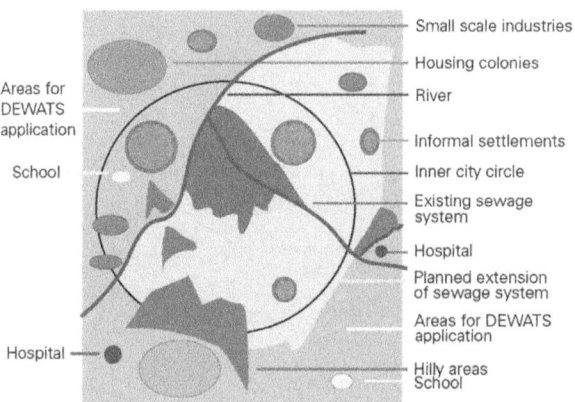

Picture 4_9: Example of how a sanitation map, detecting areas suitable to centralised and decentralised wastewater-treatment solutions might look

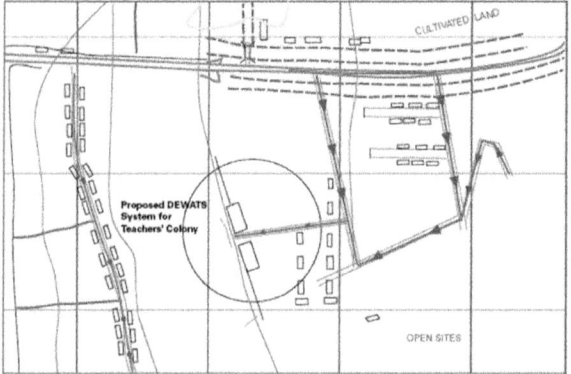

Picture 4_10: GIS-optimised positioning of decentralised sewage and wastewater-treatment facilities within a housing scheme Karnataka, India

4.4 Financial analysis

4.4.1 Comparative cost analysis for infrastructure development

Economic parameters have a major influence on technology selection. Available funds must be allocated in such a way that the required treatment efficiency is met, while being cost-efficient in providing treatment of the desired quantity of wastewater.

Centralised sewage-treatment systems usually require high investments – not only for the treatment unit itself, but particularly for the sewerage system. Decentralised solutions, therefore, often have a comparative advantage over conventional systems, especially when they are located in dispersed settlements or serve scattered pollution points.

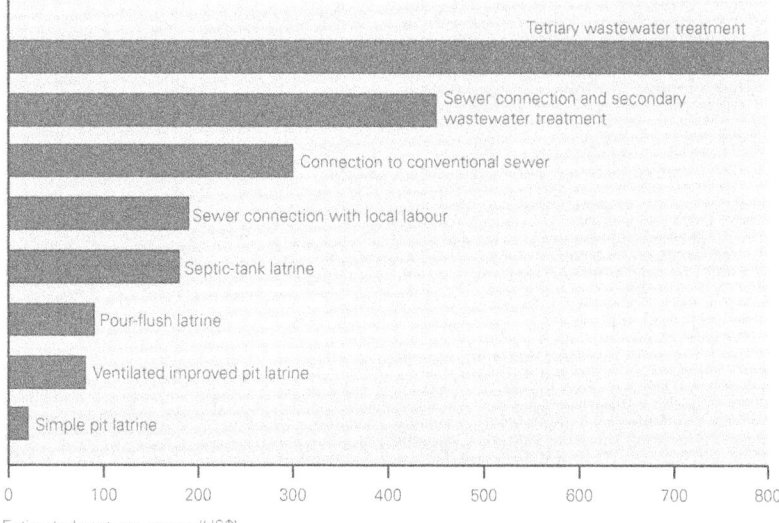

Figure 4_11:
Costs of different sanitation options.
Source: UNDP, HDR 2006

Solutions such as VIP latrines and pit latrines are at the other end of the investment scale. Their safe application, however, is usually restricted to rural areas with low groundwater levels, in order to prevent negative effects on the environment and on public health.

The highest potential of DEWATS lies in peri-urban areas. Costs for the sewerage network of a centralised system can be up to five times higher than the sewage treatment plant itself. On-site DEWATS reduce sewerage network costs significantly. Furthermore, the cost of the treatment unit should also be lower, due to a less-sophisticated technical layout.

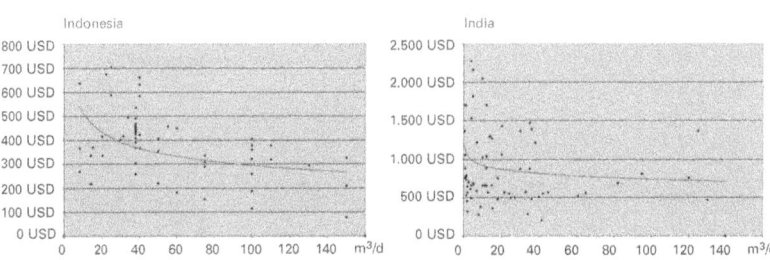

Picture 4_12:
DEWATS, Indonesia: Initial investment costs vs. daily treatment capacity in m³, 2004

Picture 4_13:
DEWATS, India: Initial investment costs vs. daily treatment capacity in m³, 2004

The exact cost of a DEWATS unit depends on the configuration of the system and the location. DEWATS are configured according to the desired treatment efficiency and various site-specific conditions. Since highest priority should be given to treatment efficiency and smooth handling of operation and maintenance, ponds rather than tanks and tanks rather than filters are recommended.[24]

24 Additional parameters, such as insect breeding, may need to be considered as well.

However, the ever-increasing value of real estate – not only in city centres, but also in fast growing peri- and semi-urban areas – eliminates treatment ponds as a viable option, due to their requirement for large surface area. Intensive treatment in compact anaerobic digesters proves more cost-effective in many locations. Due to restricted land availability, DEWATS are frequently constructed as a series of underground settlers, baffled reactors or anaerobic filters, followed by constructed wetlands and polishing ponds.

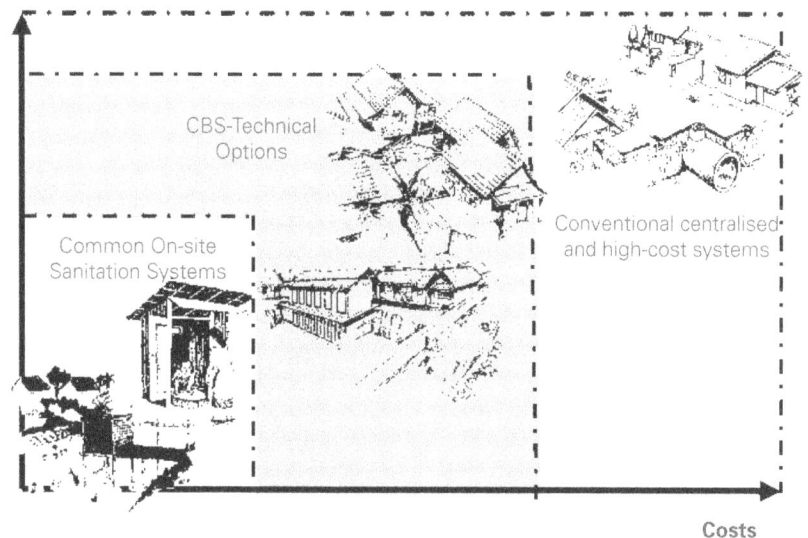

Figure 4_14:
Cost and
Convenience –
a comparison of
sanitation and
wastewater-treatment systems in
peri-urban areas

A study carried out in India and Indonesia shows the relationship between the scale of a project and the required DEWATS investment costs (land, materials and construction). The cost per volume of treated wastewater per day decreases significantly as the treatment capacity of the plant increases. The high variation of cost data within the study results mainly from varying property prices at the different locations.

In comparing different wastewater-treatment options, a comprehensive financial analysis should consider the following:
- investment in equipment and construction
- price of the land
- costs for financing
- operation and maintenance cost

4.4.2 Economic analysis in times of global warming and energy scarcity

The advantages of DEWATS over centralised systems become more apparent when external costs are included in the financial analysis. At a time of water scarcity, of rising energy prices and of global warming, decision-makers have to find their way through a multitude of important economic and ecological parameters. For example, most centralised systems rely on flush toilets, which contribute significantly to water consumption in growing urban water systems and, along with the increase in demand for clean safe water and the problem of large water losses, contribute to the deterioration of water resources. Particularly in present and future regions of water scarcity, this leads to greater water stress and higher prices for fresh-water generation. Increased water usage reduces the natural recovery capacity of water-catchment areas and, thereby, increases the cost to the national economy, as more of its
environmental assets are depleted. Furthermore, the energy need for water transport and wastewater treatment is far higher than commonly perceived.

Given the complexity of the issue, it is obvious that the discussion about sustainable energy and water use has just started. But there is evidence that such strategies have to consider the "real costs" of the use of resources. Incorporating "real costs" into water and energy prices will influence utilities and institutions in their search for the most cost-effective technological option. In particular, technologies permitting water reuse – as DEWATS do – may gain significant, comparative advantages. Addressing complex urban water systems with a more holistic view can be achieved through the framework of life cycle management (LCM). At the core is the application of environmental life cycle assessment (LCA) as one of the most important tools.[25]

25 See publications by Pillay, Friedrich & Buckley on the LCA of sanitation systems.

As mentioned earlier, wastewater management plays an important role with regard to the sustainable use of water. According to the California Energy Commission, about 4% of California's demand for electricity is for the purpose of water transport and water treatment. Though such a figure might differ significantly from region to region and country to country, this electricity demand results in an important generation of CO_2 emissions. Looking at the impact of wastes and waste treatment, analysis data from the US EPA show that in the year 2000, 4% of the world's greenhouse gas emissions are caused by methane (CH_4) and nitrous oxide (N_2O) from anthropogenic wastewater, manure and solid waste. Wastewater itself represents about 1.3% of these emissions mostly generated in ponds, septic tanks and sewer lines where the methane is not collected and burned.[26]

26 See: Perry L. McCarty: "Toward Sustainability – A Paradigm Shift in Concepts, Analyse and Goals", prese tation at WEFTEC, 2007

In order to find out how energy consumption and greenhouse gas emissions from wastewater treatment can be reduced, Perry L. McCarty did a comparative analysis of three treatment layouts in California.
- the first: a "traditional" aerobic treatment with nitrification of the excess sludge
- the second: a "traditional" aerobic treatment with nitrification followed by anaerobic digestion of excess sludge
- the third: high-rate aerobic ponds for algae production followed by anaerobic digestion of the removed algae. The further treatment steps after algae removal are: stabilisation ponds, flotation, nitrification, filtration and disinfection

In this calculation it is taken into account that
- CO_2 emissions will be penalised and
- biogas generated in anaerobic digestion is used for cogeneration of heat and power

The analysis shows, that the third layout not only produces 80% less greenhouse gas emissions than the first layout; moreover, the absence of both oxygen supply and incineration allow for an unrivalled positive energy balance (Table 6).

The study concludes that wastewater treatment alternatives need to be evaluated against climate change concerns:
- methane from wastes must be contained
- desired alternatives are those that reduce both greenhouse gas emissions and power consumption
- anaerobic digestion is likely to be an attractive component of the alternatives and
- wastewater is considered a resource for water, energy and plant nutrients

Figures on how the energy consumption and greenhouse gas emissions from production of manufactured mineral fertiliser and its transport over long distances can be prevented through recovery of nutrients from wastewater is given in the SuSanA factsheet "Links between sanitation, climate change and renewable energies" (Sustainable Sanitation Alliance, working group 3).

As shown in section 3.3 (good practice), DEWATS solutions can serve as a model case for sustainability coming in response to climate change concerns:
- securing access to basic sanitation services,
- protecting natural resources and
- providing opportunities for reuse of water, energy and nutrients.

No.	layout	CO₂ equivalents (1,000kg/day) based on the treatment of 10,000kg BOD₅/day								energy costs – US$ 1000/year treatment of 10,000kg BOD₅/day			
		BOD re-moval	incine-ration	digestion CO_2	CH_4 oxi-dation	CH_4 losss [1%]	nitrifica-tion	energy usage	Total	oxygen supply	CO_2 penalty [US$20/tonC]	excess power	Total
1	Aerobic + Incineration	3.6	20.4				2.8	2.2	29	178	58		236
2	Aerobic + Digestion	3.6		2.6	5.4	1.1	2.8	(3.3)	12.2	178	24	(299)	(97)
3	Algae + Digestion			3.4	6.8	1.4	1.2	(6.6)	6.2		12	(378)	(366)

Table 6:
Greenhouse gas emissions and energy consumption from wastewater treatment in Calif
Source: Perry L. McCarty, 2007[26]

Picture: 4_15:
Timbuktu, Mali. Not only in semi-arid regions: flush based toilet systems are often economically and ecologically unsustainable

4.4.3 Economic considerations for point-source polluters

Regional or urban planners may apply different economic-decision criteria to those of the owners of hospitals, small or medium enterprises, or residential estates, who are urged to find an efficient and cost-effective wastewater-treatment solution. While planners focus on the long-term overall development of a region, those running institutions or businesses are concerned with the compliance of legal-discharge standards, often at short notice. In these cases, decentralised wastewater-treatment solutions are frequently the only option; so the most appropriate decentralised option must be chosen. Decentralised treatment can be provided by:
- rotating, biological disc reactors
- trickling filters
- activated sludge processes
- fluidised bed reactors
- sequencing batch reactors
- or DEWATS, as described in this book

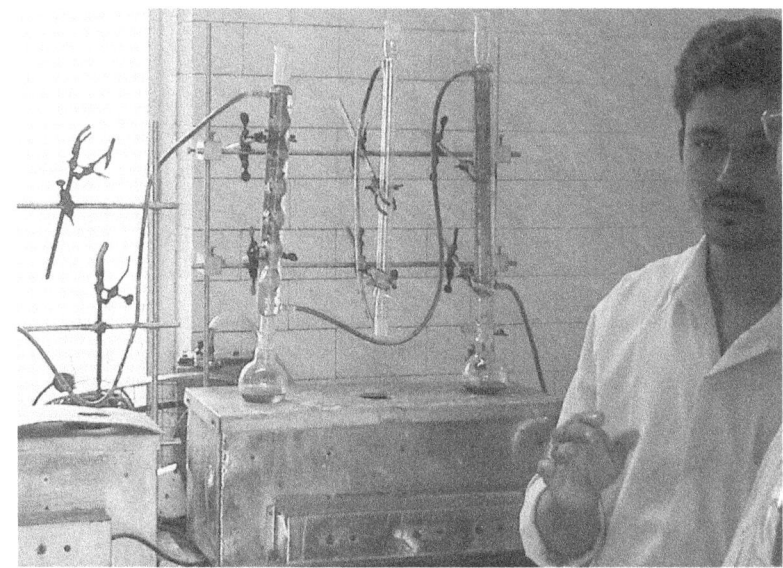

Figure 4_16
Many high-tech wastewater-treatment systems function inefficiently because it's not possible to have qualified staff to operate and maintain them

Since DEWATS are based on simple technology, which requires minimal operation and maintenance, they are favourable with regard to investment and running costs. Other technologies may require continuous support by qualified staff – often this is neither available nor affordable.

In theory, sound economic analysis requires comparable data about the various systems to be compared. In reality, the specific site conditions and the priorities of the decision-makers prevent the formulation of a standardised comparison and decision process. Every site requires its own assessment.

At the very least, the following parameters should be considered:
- potential for reduction of wastewater quantity
- potential for reduction of pollution load
- geography, geology and topography
- space availability
- availability of qualified staff for the required tasks
- discharge standards
- social environment and neighbourhood

Depending on the situation, the final decision usually has a strong socio-economic bias:

> "It has been shown that, under certain local circumstances, large variations in economy are to be expected, but the general conclusion (...) is that the economy of the various treatment processes does not differ that much. In many cases the costs are approximately the same. This increases the importance of those factors which cannot be included in an economic survey. Some of these factors are limiting factors in the sense that they limit the "free" selection between the various methods. If large areas of land are not available, then oxidation ponds must be disregarded even if it is the most economically favourable solution. If electricity supply is unreliable, then activated sludge systems cannot be considered. (...) It can be argued that the factors mentioned above are purely economic in nature, e.g. a reliable electricity supply is merely (!) a matter of economy. However, the costs involved in changing these factors to non-limiting factors are so high that there is no point in including such considerations here."[27]

[27] The Danish Academy of Technical Science, industrial wastewater treatment in developing countries, 1984

Mainstreaming DEWATS – strategic planning and implementation of sustainable infrastructure

4.4.4 Parameters for economic calculation

Global estimates of return on investments in water and sanitation published by WHO and the World Bank show that the return on a US$1 investment is in the order of US$ 5-34, depending on the intervention.

The benefits reflect a range of expected financial and economic savings to the intervention beneficiaries, including time savings due to easier access, gain in productive time and reduced health care costs saved due to less illness, and prevented deaths. The results are impressive and provide evidence that all water and sanitation improvements are cost-beneficial in all developing world sub-regions.

While investments in water and sanitation today are recognized as highly cost-effective interventions contributing to all Millennium Development Goals, it has to be added that the above mentioned studies are based on social and not financial cost-benefit analysis: costs reflect mainly financial costs whereas economic benefit is measured in terms of public health and social welfare (focussing on a real but hypothetical set of benefits) and not financially measurable benefits.

In other words: wastewater-treatment systems are not implemented to generate income. Although valuable by-products are created, such as biogas as a renewable-energy source, sludge as an organic fertiliser, or recycled water for the reduction of overall fresh-water consumption, wastewater-treatment systems are primarily infrastructural services, which must be financed by public/private bodies or individuals.

As the price of natural resources, such as oil or phosphorus, continues to increase, the valuable by-products of wastewater-treatment units will begin to play a greater role. In most cases, these products currently do not generate

Picture 4_17:
The integration of DEWATS into the infrastructure – here is a parking lot in Java – can reduce investment costs

enough return to reach a financial break-even point. However, new macro-economic tools, like regulations that promote electricity supply to the grid and new power-generating technologies, are beginning to affect the market. Significant returns should be possible at sites with intensive animal husbandry in the very near future. As a general rule of thumb, however, classical financial cost-benefit analysis does not fit the economy of wastewater treatment – yet.

The annual cost method appears to be a more apt economic indicator. It creates a more comprehensive picture of the economic implications by factoring depreciation on capital investment and operational costs into the calculation. Expenses to the polluter, like discharge fees, or income from the reuse of by-products are analysed on an annual basis. A spreadsheet for computerised calculations is presented in chapter 10.

Cost of land

Data about the cost of land may be essential when comparing different treatment systems. The applicability of sand filters or ponds is affected more by the price of land than the applicability of compact anaerobic digesters; where land prices are high, compact tanks – not ponds or filters – will be the natural choice. The value of real estate can vary widely, depending on the location. In some cases, it may contribute up to 80% of the investment cost.

Construction costs

Annual costs are influenced by the lifetime of the hardware. It may be assumed that the building and ground structures have a lifetime of 20 years, while filter media, most pipelines, manhole covers, etc. are only likely to last for 10 years. Other equipment, such as valves, gas pipes, etc., remains durable for six years. All structural elements should be categorised into one of these three categories. It is assumed

Picture 4_18:
DEWATS are qual products – plannin and monitoring of implementation must be carried o by experienced st

that full planning costs will reoccur at the end of the lifetime of the main structure, i.e. in about 20 years. In any individual case, the costs of planning can be estimated. Costing will be carried out by an experienced local engineering team being also responsible for designing and supervising the implementation of DEWATS used for. Due to the high-quality requirements of decentralised systems, engineering costs are likely to be relatively high. In addition, other labour costs plus laboratory costs for the initial testing of unknown wastewaters must also be included.

	Items of work	Unit	Quantity
1	Earth work, excavation for baffled reactor	m³	478.40
2	Plain cement concrete (PCC), 10cm thickness, floor of the tank	m³	23.92
3	Sand filling, 10cm thickness	m³	23.92
4	Reinforced cement concrete (RCC), vertical slabs for outer walls, internal baffle walls	m³	70.65
5	RCC, cover slab, 15cm thickness	m³	23.23
6	Plastering inside the baffled reactor using 1:4 mortar	m²	1,078.00
7	Pre-cast ferrocement baffle walls, 3cm thickness with necessary brick pins	m³	236.87
8	Supplying and fixing 6-inch pipes for inlet & outlet	m	12.00
9	Supplying and fixing 6-inch T-pipes	no	4.00
10	Filter media for anaerobic filters	m³	19.50
11	Manhole, size: 450mm x 450mm	no	15.00
12	Manhole, size: 600mm x 600mm	no	2.00
13	Filter drains for reusing treated water for irrigation	m	200.00

Table 7: Materials required to construct DEWATS to treat the wastewater of approximately 1,000 people (sewage production per day 80m³/d)

Running costs

Running expenses include the cost of personnel for operation, maintenance and management, including monitoring. Cost is based on the amount of time needed for qualified staff (including staff trained on the job) to attend to the plant. The time required for plant operation is normally assessed on a weekly basis. If inspection and attendance are covered by permanent staff, cost calculation is simple. Special services, requiring external work force, incur additional costs. Shared facilities, created by attaching 5 to 10 households to one DEWATS, are likely to be 10% cheaper than individual plants. In such a case, operational responsibility must be clearly defined to ensure reliable maintenance and sustainable operation.

Open systems, such as ponds or constructed wetlands, require more regular attendance than closed systems, as they may be damaged or disturbed by animals, stormy weather or falling leaves. The cost for desludging and sludge treatment, however, will be higher for heavily loaded tanks than for ponds, which receive only pre-treated wastewater.

Picture 4_19:
Service vehicle in Vietnam – desludging costs are important for financial analysis

Additional benefits from wastewater treatment

The market value of wastewater treatment by-products can be estimated by calculating the price of the products that they substitute.
Especially in dry regions, water is a major cost factor for consumers. Recycling DEWATS-treated wastewater, therefore, may considerably reduce the water bills of private consumers, companies and other entities. As described in section 3.3 "good practices", the Aravind Eye Hospital in Pondicherry, India reuses 307m³/d of its treated wastewater for gardening or toilet flushing purposes, while the Bangalore-based Alternative Food Ltd. feeds a major part of its daily treated 30m³/d of wastewater back to its production processes.

Biogas has an economic value as a renewable-energy source, which can substitute other fuels. Approximately 200 litres of usable biogas are produced per kilogram of removed COD. The actual gas production equals 350 litres of methane (500 litres of biogas) per kilogram total BOD; however, a part of the biogas remains dissolved in water, especially at low wastewater strength. Biogas contains 60 to 70% methane. One cubic metre of methane is equivalent to approximately 0.85 litres of kerosene.

To allow biogas utilisation, the structure must be gas-tight and additional volume must be provided for storage. Pipes and valves are required to transport the gas to the place of consumption. The cost of operational and maintenance attendance is likely to be approximately 50% higher if biogas is used. Further additional investments enabling the use of biogas include, approximately, 5% to the cost of long-lasting structures (20 years, lifetime), another 30% to the cost of internal structures (10 years, lifetime) and an additional 100% of the cost of equipment (6 years, lifetime). The finance costs of the additional investment must also be considered.

If the use of biogas proves to be too costly or complicated, capturing and flaring (direct burning without use) should be considered for environmental reasons: methane is a greenhouse gas with a high global warming potential.

Figure 4_20: Aravind Eye Hospital reduces its water bill significantly by reusing treated wastewater

Figure 4_21: Modern rice boiler – the use of biogas is not only of interest as a way of creating return; it should be mandatory for ecological reasons

Treated wastewater can also be used to generate income through agricultural production or fish-farming. Safety issues are treated in section 11.4 "Reuse of wastewater and sludge" (page 318). Knowledge about the size and management of the farm, as well as an assessment of the market for selling the products, will assist in making economic predictions. Experience has shown, however, that exact predictions are difficult to make.

Capital costs

If investment capital is borrowed from a bank, direct capital costs – in the form of interest – must be paid. On the contrary, if one's own money is invested, the cost of this capital is indirect because it could be used in other profitable ways (purchase of raw material for production, investment in shares or bank deposits, etc.).

For calculation purposes, annual capital costs of 8 to 15% of the investment can be assumed, depending on location and current economic-market developments.

Picture 4_22:
Floriculture in China – slurry fr[om] bio-digesters is a resource for org[anic] farming

4.4.5 Sustainable financing schemes for sanitation programmes – multi-source financing and willingness to pay

At point-source discharges, such as small and medium enterprises, hospitals, etc., wastewater-treatment may be financed exclusively by the polluter – with or without subsidies or credit lines. Generally, however, sanitation and wastewater treatment must be viewed as a service provision, similar to water and electricity supply. Comprehensive analysis of local conditions is necessary to develop reliable financing schemes for residential areas.

The economic situation of the users plays a large role in the determination of applicable financing schemes. In industrialised countries, sanitation services are in most cases, paid for by the users themselves (in-house toilets are paid for directly, sewage lines and treatment systems are paid for through user-fees and tax systems). In most developing countries, however, large sections of the population cannot afford to participate in a full-cost coverage system. So the question arises: to what extent are users able to participate financially, and what alternative cost-recovery systems can be applied?

The World Bank promotes the following financing schemes for improvements in the sanitation sector:
- "Households pay the bulk of the cost incurred in providing on-site facilities, including on-site sewer connections
- Residents of a block collectively pay the additional cost incurred in collecting wastes from individual houses and transporting these to the boundary of the block
- Residents of a neighbourhood collectively pay the additional cost incurred in collecting wastes from blocks and transporting these to the boundary of the neighbourhood
- Residents of a city collectively pay the additional cost incurred in collecting wastes from blocks and neighbourhoods and transporting these to the boundary of the city or treating it in the city"[28]

Source:
www.irc.nl/
page/6456

Full-cost coverage should be achieved in residential areas with higher income levels. In poorer areas this cannot be expected, as surveys frequently indicate that "sanitation" is rather low on residents' priority lists for spending. At the same time, 100 per cent of charity driven approaches have failed repeatedly in the past; A substantial contributions from users, therefore, is perceived as an indicator for community appreciation of the project and should be considered a "must" for successful sanitation programmes, even if the contribution covers only a small fraction of the total cost. No sanitation activities without a substantial contribution by users!!

The local situation and relevant financial-boundary conditions of all stakeholders must be assessed to determine the appropriate contribution levels for the poorer members of the population. In dealing with sanitation issues, public decision-makers must achieve balance between social-equity issues and their financial constraints. Sanitation and wastewater-treatment services can be provided through multi-source financing, based on recovering costs from users and from public sources from local, regional and central governments and/or international donor organisations.

Alternatively, good experience has also been gathered in projects, where well-off areas cross-subsidised their poorer counterparts. No matter which approach is favoured, financial schemes should always focus on the long-term objective, to ensure sustainable operation of the sanitation and wastewater-treatment systems.

The following elements are essential in the development of a financial scheme for a sanitation programme:
- assessment of available public and private funds, users' economic status, willingness to pay, etc.
- technical feasibility study – identification and analysis of different layouts for sanitation and wastewater-treatment facilities
- calculation of overall project costs, including operation and maintenance – based on experience from pilot projects and/or preliminary tendering
- informed-choice assessment of different long-term, multi-source financing schemes – resulting in development of financing mechanisms and definition of user fees

Sound financial planning must take not only the initial investments into account, but also the long-term costs of continuous operation!

Sanitation programmes must gain the acceptance of the user; without user acceptance any financial scheme will fail. Users must express a definitive "willingness to pay" to guarantee sustainability. Experience shows that "willingness to pay" is often restricted to amounts, perceived by the user as benefiting them and in line with their priorities. Public health benefits like reduction in medical cost and lost working time do not necessarily rank very high among these. In many cases, studies to determine the "willingness to pay" show that users in weak economic situations are not willing and/or not in the position to pay for wastewater treatment (sewage systems and wastewater-treatment units). Under these circumstances, user fees covering the cost for operation and maintenance services can be considered a substantial and acceptable contribution.

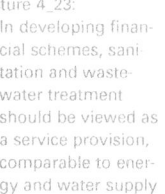

Figure 4_23:
In developing financial schemes, sanitation and wastewater treatment should be viewed as a service provision, comparable to energy and water supply

Figure 4_24:
Residential areas with high income levels should achieve full-cost coverage for sanitation and ideally assist in cross-subsidising poorer areas

Picture 4_25:
Community Sanitation Centre in Bali, Indonesia – unlike water and energy supply, sanitation service provision has few mechanisms for making users pay

Besides this problem, providers of sanitation and wastewater-treatment services face an additional challenge: unlike water and electricity supply, the service cannot be cut off if users refuse to pay. Once sanitation and wastewater equipment has been installed, few sanction mechanisms exist; users will find other ways to dispose of their waste – with adverse public health consequences for the whole community.

CBS programme planning and implementation

The profile of each CBS (Community-Based Sanitation) programme has to be country, site and situation specific. Nevertheless, in this chapter we will introduce the core elements of successful CBS implementation. The outlined programme-implementation steps are based on the project experience of "good practice" examples and guide the reader through his or her own programme and project development.

The institutional background has a significant impact on programme initiation. While organisations experienced in infrastructural development in poor areas might be able to develop institutional capacities fairly rapidly, other organisations might depend on the collaboration with other institutional players. In such a case, the greatest challenge will be to streamline the process and contributions of all partners.

The goal of any sanitation programme should be long-term sustainability with maximum positive impact. From the preliminary needs assessment in the very early stage of a programme, up to the disposal and treatment of sludge, a multitude of tasks have to be completed. The efficient setting-up and implementation of such a programme requires early identification of the different necessary tasks and who is responsible for carrying them out.

5.1 Stakeholders in CBS programmes

Sustainable infrastructure development and sanitation programmes must coordinate and streamline a multitude of stakeholders and resources. The active participation of different parties should span the entire development process, from the preparation phase, to planning, implementation, monitoring, and final evaluation. Participation improves the sustainability and performance of the project. Ownership ensures stakeholder commitment and participation, thereby reducing supervision costs.

Efficient, cost-effective and sustainable implementation requires systematic involvement of different stakeholder groups:
- Primary stakeholders – residents and direct users of the implemented measures
- Secondary stakeholders – groups with a direct or indirect responsibility in the programme. These include the leading agencies (public, NGOs, etc.), planning authorities, and health and environmental departments
- Tertiary stakeholders – providers of special services for construction, maintenance and sludge management

5.2 Responding to basic needs – active involvement of beneficiaries and residents

CBS programmes respond to the needs of residents in a given area. In most cases, the programmes target residents of poorer areas to provide them with improved in-house toilets or with additional sanitation services, such as toilets, showers or washrooms in Community Sanitation Centres.

The active involvement of communities in the planning and implementation process is crucial to the success of a sanitation programme because the residents:
- will use the sanitation facility – the facilities must fit their needs and practices
- have to contribute significantly to the system – financially or in kind
- may have an important role in the operation and maintenance of the sanitation and wastewater-treatment facilities

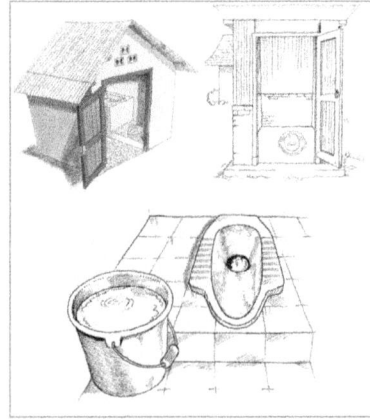

Picture 5_1: CBS programme should respond to resident need

Picture 5_2: Sanitation programmes should offer different options for improved sanitation facilities – here a pour-flush toilet

To ensure that poor residents are actively involved, the following factors are important:
- Sanitation programmes should be accompanied by health and hygiene awareness-raising campaigns
- Programme acceptance by local leaders helps to avoid unnecessary interference with social hierarchies
- Social-settlement structure and stratification, sanitation practices, informal land-holding customs, and reservations about infrastructure implementation should be understood and taken into account
- Women are often the household decision-makers with regard to domestic sanitation and sanitation practice. Therefore, they must be actively involved in determining problems, identifying underlying causes, recommending possible solutions and, ultimately, making decisions to solve the problems

Developed over recent years, "demand-responsive approaches" have become the conceptual framework of sustainable sanitation programmes. The approach treats users as clients, who express their needs, but must provide contributions in monetary terms or in kind.

Neither "demand" nor "willingness to pay" are easily measurable. Comprehensive methods have been developed to cater to users' needs: "informed choice" generates indicators for communities' and individuals' willingness to participate in the project. "Contingent valuation" (CV) provides information on potential demand and willingness to pay for different sanitation options.[29]

See: UNDP, *Willing to pay but unable to charge*, 1999

Depending on the location, "demand-responsive approaches" can result in quite different technical solutions and management configurations:
- In most parts of Eastern Java, coherent social structures mean there is a high capacity for community self-organisation and management. Decision-making processes, concerning the choice of sanitation facilities and the layout of the DEWATS, can be initiated by external facilitators and tend to run smoothly. On the whole the community manages cost recovery, operation and maintenance. Only desludging is organised by an external service provider

- In Tangerang, Indonesia, the involvement of future users in the planning process showed not only the residents' interest in improved sanitation but also in shower and laundry facilities. As the residents are mainly migrant workers, social structures are rather weak. To ensure sustainable management of the sanitation project, it was decided that BEST, a local NGO, should function as service provider. BEST ensures daily operation, maintenance and desludging of the system. The costs are covered by a fee that residents pay when they use the facility
- In Ullalu Upanagara, Bangalore, a slum with inherent social frictions, an operation and maintenance system run by the community, was set up. The Community-Based Sanitation programme was facilitated by the local NGO. Strong emphasis was put on the involvement of women in the awareness-building process for sanitation demand. One of the women groups later took over the operation and maintenance of the Sanitation Centre. Users pay a fee per use. Desludging is organised in co-operation with local government
- Although categories like "available income", "existing sanitation facilities" and "hygienic behaviour" are important parameters for a comprehensive assessment, "willingness to pay" is one of the strongest indicators. In a slum area of Mysore, India, a range of low-budget sanitation options were discussed with the residents. The potential users were only willing to pay a very small amount for the most-desired facility, a 20-toilet sanitation centre. It was only through the intervention of an experienced facilitator that it became clear that the residents had a long-term vision – for each house to be supplied with an in-house toilet. Although this was a very expensive option, they were prepared to contribute much more towards this solution. The sanitation centre would probably not have been acceptable in the long term

CBS programme planning and implementation

The following suggestions may help successful collaboration in poorer residential areas:
- Residents must contribute to the programme financially or in kind. However, there is no blueprint for how much must be contributed to a CBS programme. The "contribution profile" must be developed in accordance with the local social situation and the interests of the residents
- Participation needs time and resources; it is essentially a process with no guaranteed outcome. For these reasons, the financial cost of participation has to be carefully weighed against its benefits. A reasonable balance between input and output should be achieved
- CBS may interfere with the social structures of a community. Under certain conditions, participation can have a destabilising effect by creating an imbalance in existing socio-political relationships. Participatory approaches can result in conflict because existing power relationships are threatened. A sensitive approach is vital to avoid worsening the position of those who are already marginalised
- A fact-driven approach is suggested. Although the residents of poor areas must be provided with basic-needs services, an exaggerated emphatic approach may result in skewed perception of ground realities and a CBS programme with too much "wishful thinking"

Picture 5_3:
In Tangerang, residents have to pay the equivalent of 0.07 US$ for each time they use the facility. The service provider employs one operator, who ensures operation and maintenance

Picture 5_4:
In Ullalu Upanagara, local women were trained to ensure continuous operation

5.3   Local government and municipality bodies

Local public bodies play a pivotal role in successful sanitation and wastewater-treatment projects. Although the specific responsibilities of a body may vary from country to country, it is the local government (or municipality) that is usually accountable for providing sanitation and wastewater-treatment services. Furthermore, it is also responsibile for promoting health and hygiene awareness, to ensure the health of its communities and to monitor the environmental impacts.

Local government and municipalities should formulate and implement a policy addressing the sanitation backlog and water-pollution problems. They are responsible for driving the local processes set out in its policy. They must create an enabling regulatory environment through municipal by-laws – and ensure both appropriate and affordable service implementation. Furthermore, the local government must ensure that environmental standards are met, including the establishment of a system for sludge removal, treatment and disposal.

In an ideal world, local-government agencies would integrate sanitation and wastewater strategies into their local development plans and take the lead in implementing them. In the real world, different stakeholders may take the initiative to provide sanitation services, particularly to the poor. Beside public bodies, international agencies, NGOs and Community-Based Organisations may be active. In such cases, a co-ordinated strategic alliance between different stakeholders can create a greater drive for the implementation of efficient sanitation and wastewater-treatment options. Nevertheless, local government and municipalities must be fully involved.

CBS programme planning and implementation

To assist the implementation of wastewater-treatment infrastructure, the relevant bodies should:
- create a demand for sanitation improvement through health and hygiene-awareness programmes
- respond to this demand by identifying appropriate sanitation options
- prioritise these options
- integrate these working results into a planning process
- allocate funds to achieve the planned objectives
- ensure there are enough appropriately skilled people to carry out the plan
- implement the plan
- monitor and report on the results and
- ensure sustainability

Figure 5_5:
Strategic alliances between different stakeholders ease CBS programme implementation

5.4 Non-governmental organisations

A number of NGOs have launched CBS programmes and/or have become programme partners to government bodies. Which roles NGOs play within CBS programmes depends mainly on their competencies and the local situation.

NGOs can play a leading role in CBS programmes. As specialists in poverty alleviation and environmental protection, many NGOs have in-depth knowledge about the low-income groups with whom they work. They know about local sanitation practices, decision-making processes within the communities, income and expenditure patterns, and other factors, crucial to successful sanitation programme implementation.

Furthermore, many NGOs have good working relationships with the communities. So they can facilitate awareness-raising campaigns, decision-making processes or other forms of communication.

Over time, many NGOs have developed competencies as service providers. Some are active in the fields of solid-waste management, environmental counselling and/or urban planning. Other NGOs have the capacity to set up and run complete sanitation and wastewater-treatment projects, including the provision of operation and maintenance services.

5.5 Private sector

In most cases, the private sector can and should cover important CBS programme tasks. The private sector may:
- plan, design and construct sanitation infrastructure
- plan, design and construct wastewater-treatment infrastructure
- manufacture equipment
- ensure operation and maintenance of the overall scheme
- operate desludging and sludge-treatment facilities

These services may be provided on a contractual basis. Close quality monitoring of the delivered services is crucial to the sustainability of the programme.

CBS programme – detailed procedure for implementation

The success of a CBS programme depends significantly on implementing the steps in the right order. The organisation or the group of initiating bodies, taking the lead in launching a project should be aware of the complexity and usefulness of a comprehensive approach. Success depends on the co-ordinated implementation of a multitude of tasks and the integration of all stakeholders into the process.

6.1 First planning activities

An initial workshop helps to establish a common foundation between key stakeholders. Members from the leading agency (LA), NGOs – or representatives from future beneficiary groups – should be invited to form a core team. The following issues should be addressed:
- targets of the envisaged programme
- assessment of the current situation in the relevant area, regarding sanitation and wastewater
- key existing problems in sanitation, wastewater and environmental pollution
- existing experiences with relevant projects
- awareness building concerning the tasks to be fulfilled throughout the programme
- identification of relevant stakeholders to involve in the project

Picture 6_1: Stakeholder roles and responsibilities must be clearly defined at an early stage – CBS programme steps should be understood by everyone

The key programme tasks should be identified at an early stage. With these in mind, the steps of implementation should be defined to enable smooth operation. Key tasks include:
- overall programme management, including process monitoring
- developing a feasibility study
- community preparation, including health and hygiene awareness-raising campaigns
- construction
- operation and maintenance
- monitoring sanitation and environmental standards
- final sludge management

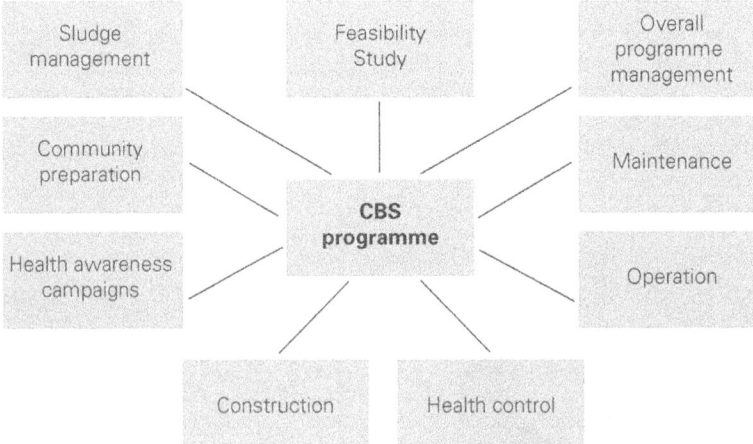

Picture 6_2:
CBS programme – core tasks

Workshop participants should identify the specific competences and resources of the various stakeholders. Their roles and main responsibilities within the programme should be assigned. The collaboration and roles of partners may vary greatly from location to location.

Furthermore, the workshop should help to establish an efficient working structure. In particular, the role of the leading agency should become clear, including its core management and monitoring responsibilities. Questions to be answered include:
- What are the tasks and responsibilities of the leading agency?
- What roles and responsibilities can the communities fulfil?
- In what areas can the private sector contribute to achieve higher quality and cost effectiveness?
- What competencies can other stakeholders, such as government agencies or NGOs, contribute?
- Which tasks cannot be fulfilled by the available stakeholders? Which measures, including staff recruitment, are needed to bridge the existing gaps?

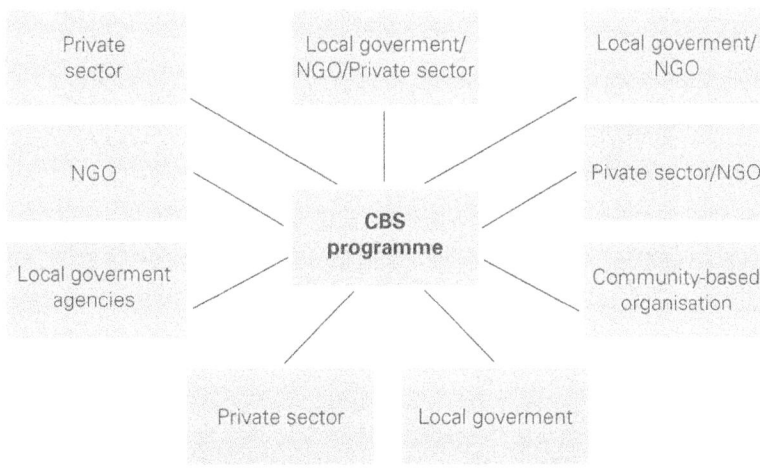

Picture 6_3:
CBS programme –
stakeholders

The signing of a contractual agreement or Memorandums of Understanding between the stakeholders at an early stage in the programme helps to establish a solid foundation for the following steps of co-operation.

6.2 The pilot project

Setting up a large programme is facilitated by a successful pilot project. Experience shows that it is better to implement a simple pilot, which can be extended, than to be too ambitious and create a complex programme that cannot be handled by the implementation body.

A pilot project should:
- clarify and strengthen the working structure of the implementing body
- provide all stakeholders with a firm understanding of the challenges and technical, social and financial requirements for an implementation programme
- develop and test appropriate instruments and tools for large-scale application
- integrate the relevant stakeholders into the implementation process
- equip executing bodies to be constructors for further activities
- create standardised procedures for the overall approach

The location of a pilot project should be representative of other local locations in the municipal area, with regard to quantity and quality of wastewater, socio-economic structure, settlement layout, etc. It should also allow for fairly smooth implementation and operation to set a positive example.

The CBS programme in Ullalu Upanagara, Bangalore, India, for example, was used to test the implementation procedure. The experience gained from the project was later used to effectively target other sites within the programme area.

A pilot project provides valuable information for future projects within the programme:
- Which feasibility-study parameters are relevant in different parts of the city?
- Which planning tools are most efficient?
- How should a demand-responsive approach look for similar target groups?
- Which DEWATS configurations would be most appropriate at similar sites?
- Which informed-choice options proved most useful?
- Which stakeholders must be involved in which implementation stages?
- Which public authorities are relevant for overall project clearance?
- Which "informal leaders" must be involved?

CBS programme – detailed procedure for implementation

- How can women be targeted through special awareness-raising campaigns?
- Which problems are likely to emerge (i.e. with land-holding)?
- Which contractual arrangements with which stakeholders proved most useful?
- How can the overall implementation process be monitored?
- What financial and in-kind contribution can be expected from the users?
- How can operation and maintenance be organised?
- What can be expected from the users and which tasks must be fulfilled by a service provider?

Evaluating a pilot project helps to optimise future planning processes.

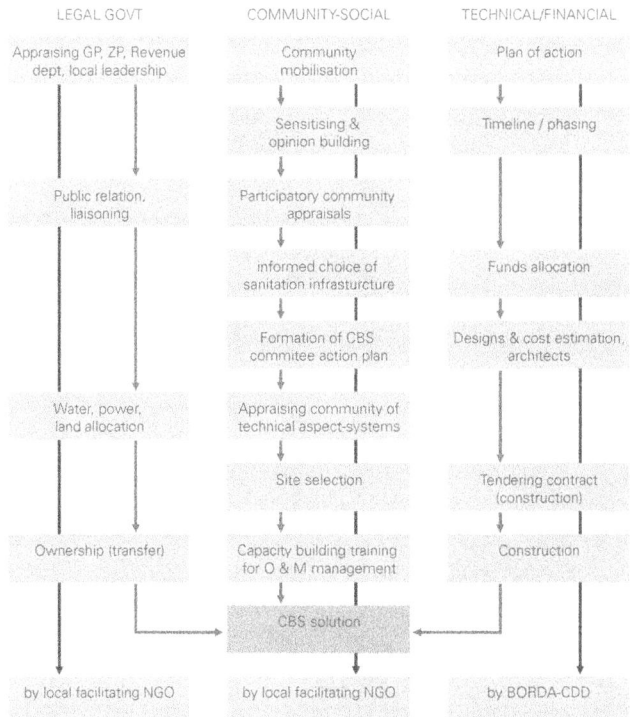

Figure 6_4:
Chart indicating the implementation process in Ullalu Upanagara
ZP = Zilla Parishad government administration on District level, GP = Gram Panchayat (village level administration)

6.3 Preparation phase

6.3.1 Kick-off workshop

Experience shows that it is beneficial to officially launch the programme with a "kick-off" workshop to which the various stakeholders are invited:
- senior government officials at the local, regional and national levels
- relevant NGOs
- representatives of the target groups
- relevant researchers
- private-sector participants
- international agencies
- media, etc.

Picture 6_5:
The kick-off workshop introduces the scope of the programme and involves senior stakeholders on various levels

The workshop should:
- communicate the results of the pilot project
- demonstrate the scope and relevance of sanitation and wastewater-treatment programmes in relation to different fields of policy and different government levels
- clarify the importance of target driven co-operation between stakeholders

The workshop is aimed at:
- creating awareness amongst decision-makers about the legal requirements, required resources and institutional backing for the programme
- developing a supportive environment – getting different stakeholders and local authorities to offer their competencies to the programme
- launching a process for the provision of financial and human resources on different government levels
- gaining support for extending the programme into other municipalities, departments, or provinces

6.3.2 Planning workshop

The planning, implementation and monitoring activities of a programme should be launched at a planning workshop. The workshop participants analyse the results of the pilot project and draw conclusions for dissemination on a larger scale. The main out-comes of the workshop might include:
- formulation of stakeholder responsibilities, timeline and resource planning
- standardisation of procedures, such as site selection, community involvement, tendering, construction, sludge management, etc.
- drafting supporting documents, such as training kits, contract forms, monitoring sheets, etc.
- formulation of capacity-building plan for key stakeholder groups

Participants of the workshop should be:
- local authorities
- NGOs
- members of the target group
- relevant private-sector participants

Ideally, the workshop results in an approval agreement between the stakeholders involved in the implementation of the programme. Discussing and agreeing on responsibilities and the stages of the project in advance helps to avoid conflict during the later stages of implementation or operation.

Picture 6_6:
 Clear agreements between project partners ease the implementation process

6.3.3 Community pre-selection and community assessment

Sanitation mapping is a powerful tool for identifying and long-listing communities eligible for sanitation and decentralised wastewater-treatment projects (see section 4.3.2, page 68). If comprehensive sanitation mapping is not possible, the long-list can be based on the experience of government authorities, NGOs or other agencies. Local authorities tend to have their own classification methodology for poor urban areas, which may be useful for the identification of potential sites.

Other criteria, which may be useful, include:
- health risks within the area
- vulnerability of the ecological system or possible environmental threats
- legal status of the settlement
- income classification.

The long-listed communities should be assessed and ranked. To enable comparison, the following information should be collected:
- current situation of sanitation and wastewater-treatment
- reports on existing sanitation programmes in comparable areas
- indicators for the community's willingness to participate in a programme
- social structure and decision-making procedures within the community
- legal status of the settlement
- land availability
- geological and topographical data

The community is provided with information about CBS and invited to a stakeholder meeting. The meeting agenda should be adapted to the specific local context. During the meeting, the CBS programme is presented and the sanitation conditions in participating communities are discussed. The contributions expected from the community must be stated clearly. Interested communities are asked to submit an expression of interest (EoI) in taking part. The EoI includes an invitation for a rapid participatory assessment (RPA).

The RPA determines if the site is suitable for DEWATS applications:
- natural gravity flow should be assured – the natural slope of the land should lead the wastewater from where it is generated to the treatment plant and then to the discharge point
- availability of water and land for construction are essential for DEWATS implementation. Illegal settlements are excluded from participation. If an area is prone to flooding, community sanitation centres are usually recommended.
- community sanitation centres require vacant land for construction. Land availability and ownership must be clear
- as sanitation and wastewater systems often have a negative image, residents living close to the planned CBS or treatment unit must agree to the chosen location. Planning procedures must take account of the time necessary for obtaining community acceptance of the facilities in their immediate neighbourhood

To make sure results are objective, representatives of all community stakeholder groups should be actively involved during the RPA process. The following RPA tools can be applied:
- ladder – assessing community willingness to contribute to the new sanitation infrastructure
- transect walk – identifying and analysing the condition of existing sanitation systems in the neighbourhood through direct observation
- problem trees – identifying and analysing community sanitation problems, their cause and effect, and whether the community intends to improve its sanitation conditions
- timeline – identifying and analysing residents' experiences with previous community-participative infrastructure projects.
- venn diagram – identifying and studying existing local community institutions, their benefits and their relationship with the community. This tool is also used to assess community readiness to operate the facility

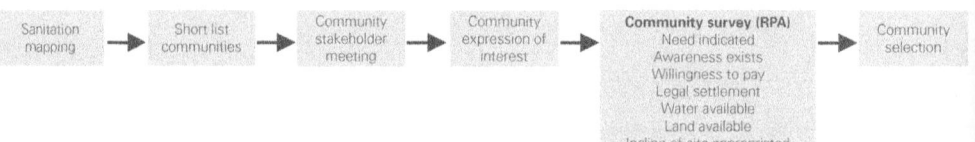

Picture 6_7:
Flow chart: from sanitation mapping to community selection

6.4 Planning phase

6.4.1 Site assessment

Successful CBS planning depends on a detailed technical survey carried out by technical experts. The relevant information can come from local government and community surveys. Government authorities and stakeholders on the ground, therfore, should be involved in the process. Some of the necessary information can also be obtained from the Rapid Participatory Assessment carried out during community selection.

Figure 6_8:
Detailed knowledge of the local situation is the basis for appropriate planning

The technical survey procedure is identical, with or without community participation. However, the process differs according to the selection of technology. Household-based sanitation systems, including simplified sewerage or off-site treatment, demand more complex surveys, planning and establishment activities than community sanitation centres with on-site treatment. The technical survey consists of four sections:

1. Assessment of general site conditions

- cartographic and topographic surveys and mapping – focused on settlement structure, topography (elevation) and site accessibility
- location and general data collection about local industries and enterprises, including home working, peripheral farming activities, restaurants and food stalls
- assessment of the number of potential users and their habits regarding water, sanitation and waste-related subjects
- survey of soil conditions at potential construction sites

2. Assessment of water and wastewater-related subjects

- assessment of water sources, including quantitative and qualitative security
- assessment of water-consumption levels of industries and households, and required quality for different uses
- assessment of domestic and industrial wastewater-generation processes, volumes, composition, discharge patterns and reuse options
- assessment of existing sanitation and wastewater-treatment systems – applied technologies, performance, responsibility for operation and management
- assessment of the rainwater-runoff infrastructure
- assessment of discharge options – survey of local water bodies with regard to quality, flow volume and location, including groundwater quality, use and level
- gathering of precipitation data for different times of the year

3. Legal background on wastewater

- gathering of wastewater-discharge standards and environmental-protection regulations

4. Building materials and tools

- assessment of local availability of building materials and tools

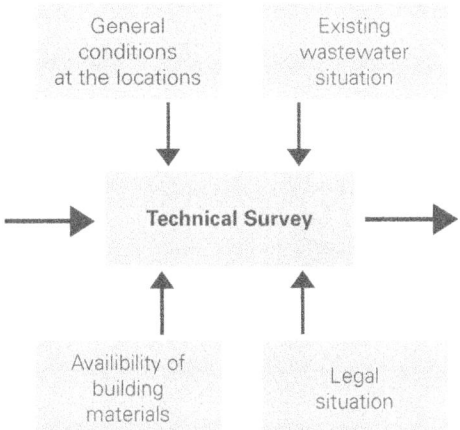

Figure 6_9:
Technical survey

6.4.2 Informed technology choice

By providing the potential users with different options for sanitation facilities and services, the principle of "more expensive systems will cost more" is communicated. The users themselves can eliminate options that do not apply to their situation.

In the informed-choice process, users learn about many possible options:
- different toilet types and layouts of toilet facilities
- different functions and layouts of community sanitation centres
- different service levels to be expected

Informed choice is usually focused on the users' preferences concerning sanitation equipment. However, the following components can also be addressed within community meetings to assess the public acceptance of their application:
- sewer layout
- treatment components
- disposal or reuse of effluent
- disposal or reuse of effluent sludge

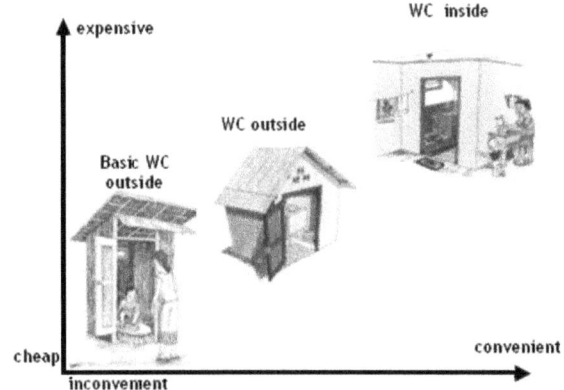

Picture 6_10:
Toilet-facilities options

Picture 6_11:
Experienced expe[rts]
should facilitate
assessments on w[il]-
lingness to pay a[nd]
informed choice –
substantial financ[ial]
contributions by t[he]
users are crucial [to]
the sustainability [of]
sanitation scheme[s]

6.4.3 Detailed engineering design

Experienced technical experts prepare the detailed engineering design applying the results obtained from the technical survey, the technology-choice discussions and the assessment of other local factors. The solutions should be discussed with local decision-makers and community leaders – to identify potential problems – at different stages of the design process.

The superstructure of CSCs *can* be constructed with standardised designs. But, public acceptance will increase significantly if the users get to have their say. The number of toilets, showers, water points and laundry places is calculated in accordance with the estimated number of users.

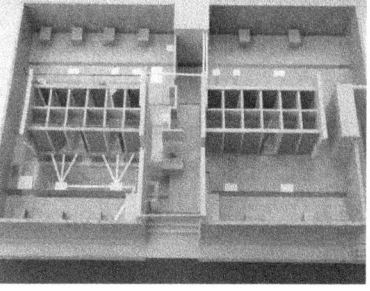
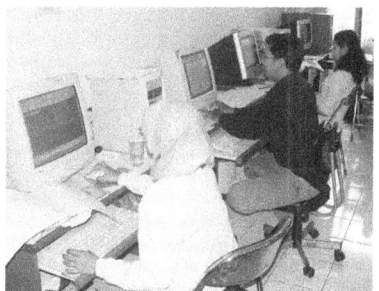

Picture 6_12:
The design should take into account the specific local conditions and the super preferences

Picture 6_13:
Experienced technical experts prepare the design

For in-house sanitation, the appropriate location for toilets, showers and washbasins is determined together with the users. The sewerage system, including inspection chambers, must be designed according to the flow volume, peak flows and slope.

The most appropriate discharge or reuse option is selected and designed in accordance with possible applications, the local surroundings, the legal situation – and the treatment efficiency of the chosen technology.

The connection of small home-industries (e.g. tofu production), restaurants or food stalls can have a strong impact on the performance of the system. Where a simplified sewerage system with off-site treatment is constructed, special attention must be paid, therefore, to commercial wastewater. If additional inflow of such wastewater is likely in the future, this should also be considered in the plant design.

Innovative designs can reduce the cost of the facility. Examples include:
- baffled reactors and anaerobic filters located under pavements, carparks, playgrounds or streets
- positioning the facility to minimise land use and length of sewerage systems
- application of reliable standards to minimise sewer diameters

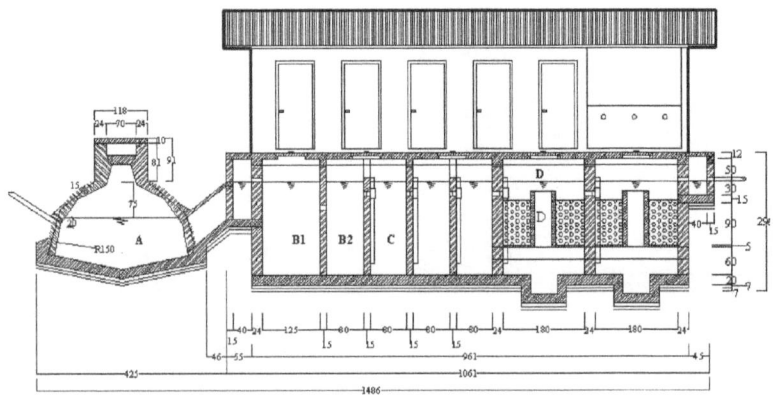

Picture 6_14: Detailed engineer design

6.4.4 Economic planning

Economic planning includes calculating the overall project costs and developing a strategy for covering these costs.

The costs to be considered are for:
- land
- materials
- labour
- supervision – including optional planning
- operation – electricity, water, service provision etc.
- maintenance – desludging and sludge treatment.

Where Community Sanitation Centres (CSCs) and DEWATS are to be constructed, the question of land ownership must be resolved. In order to purchase real estate, negotiations must be held with the owner – a site is either private or municipal property. If the owner is not willing to sell the land, perhaps a long-term lease of 15 to 20 years can be obtained. Alternatively, a usufruct may be granted for publicly owned sites. Agreements on the land ownership or renting scheme should be finalised before more activities start.

The quantity and volume of the necessary building materials can be calculated using the detailed engineering design. The total costs for construction depend on the local context:

A local government makes a financial contribution to a project, market prices may not apply; many authorities have price codes, which must be used instead

- if construction is carried out by employees, local wages and material costs apply;[30] in-kind community contributions should also be included in the calculation
- if there is a bidding process, estimated prices should be used – their final prices factored in after the tender is accepted
- if operation and maintenance is supplied by a service provider, a call for tenders can be launched
- if operation is carried out by the community itself, local labour and running costs should be added in

Cost coverage can be achieved with possible financial contributions from:
- residents
- public authorities and
- international donors

The fee structure for a CSC should be based on the completed assessment of the users' willingness to pay. If full-cost coverage cannot be expected, further implementation should be postponed.

6.4.5 Agreement on implementation and landholding

In some locations it was helpful to formalise the overall process by signing a Memorandum of Understanding (MoU). The document, signed by a legal representative of the community and relevant municipal bodies, became a central part of the CBS-development plan. It is very important to ensure that the whole community supports the project, as individual disagreements can jeopardise the entire process at a later stage. A community meeting should be held, therefore, to discuss the results of the project planning and to smooth out any remaining concerns before commencing with implementation.

The main points of the MoU included in the development plan are:
- geographical and topographical maps
- detailed engineering design
- budget plan – including the schedule for disbursements and detailed statements about the contributions from different stakeholders
- implementation schedule
- operation and maintenance plans
- ownership
- responsibilities during implementation (planning & construction)
- responsibilities after implementation (operation & maintenance)

6.5 Implementation phase

6.5.1 Task planning

The implementation schedule is developed from the tasks and respective workloads defined during the economic-planning phase. Tasks are grouped into categories and listed on a spreadsheet.

No.	Tasks	Quantity	Unit
A. Wastewater-treatment system tasks			
1	Prepare building site & levelling	1.00	–
2	Provide office space and storage	1.00	–
3	Documentation	3.00	–
1	Levelling	64.43	m^3
2	Prepare sand-bed	12.00	m^3
3	Refill earth	16.11	m^3
1	Lay brickwork	77.93	m^3
2	Prepare concrete reinforcement	40.64	m^3
3	Plaster	124.27	m^2
4	Prepare working subgrade	6.00	m^3
5	Lay brickwork	560.32	m^2
1	Pipes PVC Ø 6", l = 20 feet (~6m)	2.00	pieces
2	Pipes PVC Ø 4", l = 20 feet (~6m)	61.00	pieces
3	Pipes PVC Ø 2", l = 20 feet (~6m)	2.00	pieces
4	T-piece	80.00	pieces
5	Manhole cover	13.00	pieces
B. Sewerage system tasks			
1	Exavation	305.00	m^3
2	Refill earth	101.67	m^3
3	Prepare sand-bed	75.50	m^3
1	Prepare concrete reinforcement	0.94	m^3
2	Watertight plastering of manholes	0.00	m^2
3	Open road surface	398.00	m^2
1	Pipes PVC Ø 6", l = 20 feet (~6m)	152.50	pieces
2	Pipes PVC Ø 4", l = 20 feet (~6m)	48.00	pieces
3	Mount prefabricated manhole	35.00	pieces

Table 8:
Example of an implementation schedule

The amount of work predicted in the detailed engineering design and economic planning is entered into the table. Task sequences are determined and the required time for each task is estimated.

Time planning requires CBS project experience and knowledge of local conditions (e.g. weather conditions during different seasons, cultural and religious events and holidays, financial allocation patterns of governments and the community). The schedule must be prepared, therefore, by experienced local staff, or by experts in co-operation with people who have local knowledge.

The average time required to construct a community sanitation centre in Indonesia ranges between 70 and 90 days. For household-based sanitation systems, including simplified sewerage and off-site treatment, 90 to 110 days can be estimated.

Picture 6_15:
Task planning should be carried out by experts

6.5.2 Quality management

The requirements for a sound planning and implementation are too often underestimated. Quality management during construction, therfore, is an essential element of the successful long-term operation of CSC/DEWATS. The systems must be constructed as high-quality products. Poor construction quality and minor faults, such as bad plastering or the use of poor-quality bricks, may result in the failure of the entire system. Construction should only be carried out, therefore, by contractors and companies who can be guaranteed to use of high-quality materials and labour. Different quality-control models have proven successful in the construction of anaerobic wastewater-treatment systems:

- In Nepal, only licensed contractors are entitled to construct biogas plants. If the qualitiy of work is unacceptable, the constructor risks losing his/her licence and will be excluded from the programme.
- In Indonesia, a network of NGOs promoting DEWATS and CBS has developed an internal certification system to assure the proper application of new quality standards. Only certified products and personnel may participate in the implementation of such facilities. High standards are ensured by certified:
  - planners
  - foremen
  - site engineers
  - supervisors
  - design engineers and
  - senior design engineers[31]

Picture 6_16:
Ensuring good-quality workmanship is essential for a successful CBS programme

Requirements include:
1) education: minimum level or experience,
2) training – completion of a training programme
3) examination – training combined with tests,
4) repeated examination – every second year,
5) practical experience – adequate involvement in the different steps of DEWATS implementation.

Efficient quality management goes hand in hand with capacity building; on-site training measures should be an integral part of any programme. Especially in programmes aimed at large-scale implementation, quality-control and standardisation procedures must become common elements to ensure effective and efficient use of the resources.

6.5.3 Construction

The construction process and its main components are summarised in Table 9:

	Community sanitation centre	Simplified sewerage, incl. off-site treatment	Shared septic tank
Sanitation module	**Preparation work** • Survey and prepare site • Arrange materials procurement • Arrange tools and machinery • Arrange work force **Earth work** • Excavate • Levell • Fill earth and compact • Fill sand and compact **Concrete work** • Cast concrete slab • Carry out brick work • Plastering **Carpentry & roofing** • Topping-out the truss • Roofing **Assembly work** • Mount piping • Sanitary equipment • Water & electricity supply • Other interior fittings	**Preparation work** • Survey and prepare site • Arrange materials procurement • Arrange tools and machinery • Arrange work force Install sanitary equipment at appropriate location **Concrete work** • Cast concrete slab • Carry out brickwork • Plastering Install and connect piping to collection system **Assembly work** • Mount piping • Sanitary equipment • Water supply • Other interior fittings	**Preparation work** • Survey and prepare site • Arrange materials procurement • Arrange tools and machinery • Arrange workforce Install sanitary equipment at appropriate place **Concrete work** • Cast concrete slab • Carry out brickwork • Plastering Install and connect piping to collection system **Assembly work** • Mount piping • Sanitary equipment • Water supply • Other interior fittings
Collection module	Install and connect piping to treatment system (sub-soil below CSC)	**Preparation work** • Survey and prepare location line • Arrange materials procurement • Arrange tools and machinery • Arrange workforce **Earth work** • Excavate • Levell • Fill earth and compact • Fill sand and compact **Install inspection chamber** • Cast concrete slab • Carry out brickwork • Plastering Lay sewer pipes Fill construction ditches and compact	**Preparation work** • Survey and prepare location line • Arrange material procurement • Arrange tools and machinery • Arrange work force **Earth work** • Excavate • Levell • Fill earth and compact • Fill sand and compact **Install inspection chamber** • Cast concrete slab • Carry out brickwork • Plastering Lay sewer pipes Fill construction ditches and compact

Table 9: Detailed description of construction process

CBS programme – detailed procedure for implementation

	Community sanitation centre	Simplified sewerage, incl. off-site treatment	Shared septic tank
Treatment module	Preparation work • Survey and prepare site • Arrange materials procurement • Arrange tools and machinery • Arrange workforce Earth work • Excavate • Levell • Fill earth and compact • Fill sand and compact Concrete work • Cast concrete slab • Carry out brickwork • Plastering Assembly work • Mount piping • Fill filters	Preparation work • Survey and prepare site • Arrange materials procurement • Arrange tools and machinery • Arrange workforce Earth work • Excavate • Levell • Fill earth and compact • Fill sand and compact Concrete work • Cast concrete slab • Carry out brickwork • Plastering Assembly work • Mount piping • Fill filters	Preparation work • Survey and prepare site • Arrange materials procurement • Arrange tools and machinery • Arrange workforce Earth work • Excavate • Levell • Fill earth and compact • Fill sand and compact Concrete work • Cast concrete slab • Carry out brickwork • Plastering Assembly work • Mount piping • Fill filters
Discharge module	Preparation work • Survey and prepare location • Arrange materials procurement • Arrange tools and machinery • Arrange workforce Earth work • Excavate • Levell • Fill earth and compact • Fill sand and compact Install inspection chamber • Cast concrete slab • Carry out brickwork • Plastering Lay sewer pipes Fill construction ditches and compact	Preparation work • Survey and prepare location • Arrange materials procurement • Arrange tools and machinery • Arrange workforce Earth work • Excavate • Levell • Fill earth and compact • Fill sand and compact Install inspection chamber • Cast concrete slab • Carry out brickwork • Plastering Lay sewer pipes Fill construction ditches and compact	Preparation work • Survey and prepare location • Arrange materials procurement • Arrange tools and machinery • Arrange workforce Earth work • Excavate • Levell • Fill earth and compact • Fill sand and compact Install inspection chamber • Cast concrete slab • Carry out brickwork • Plastering Lay sewer pipes Fill construction ditches and compact

6.5.4 Pre-commissioning test

In order to ensure good construction quality, the system is tested upon completion by technical experts from the local authority. All technical modules are evaluated with regard to engineering design, quality of workmanship and functional efficiency. System flow and the water-tightness of the piping and treatment system are scrutinised closely. Technical drawings and checklists serve as tools for the pre-commissioning test.

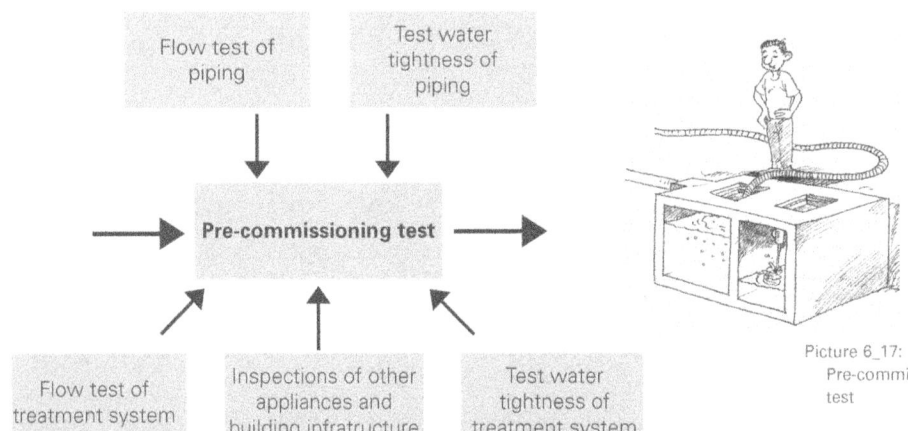

Picture 6_17: Pre-commission test

Picture 6_18: Only fully functioning systems are accepted

6.5.5 Parallel training measures

- Environmental-health training is encouraged in all CBS projects. The training is targeted at everyone in the community – all members and individuals involved in operational activities. The aim of the training is to explain the importance of sanitation facilities for personal and environmental health and to provide an understanding of the broader context of sanitation. The main subjects include personal hygiene, handling human excreta and rubbish, disease transmission and background information on particular diseases, such as diarrhoea, typhoid and dengue fever. Training is based on the guidelines of the PHAST (Participatory Hygiene and Sanitation Transformation) Initiative, jointly developed by WHO and UNDP/World Bank Water and Sanitation Programme.

- Function and application training is a basic training module for all users of the new sanitation infrastructure. The aim is to impart basic knowledge about how the system works. Correct use of the system is explained along with information about what may harm functional efficiency. The training comprises a theory and a practice module. The practical training should be carried out on site after construction is finished.
- Operation and maintenance training is only given to people directly involved in the operation and maintenance activities. Basic information on wastewater treatment and the function of the system is provided as well as an overview of all routine tasks for operation and maintenance. The various operational faults and necessary maintenance steps are also discussed.

6.6 Operation phase

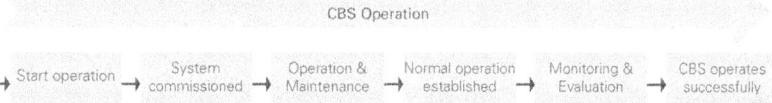

Figure 6_19: Flow chart: operation phase

6.6.1 Start operation

Operation is usually initiated by technical experts from the local authorities in collaboration with the intended operators. For fast start-up, baffled reactors and biogas plants should be inoculated with digested sludge from existing anaerobic wastewater-treatment units, such as septic tanks. After starting the system, operators are briefed on operation and maintenance.

The system formally starts operation at a hand-over ceremony with the community and/or the operating agency. Particularly in poor areas, such events should be perceived as a positive gesture towards the development of the area.

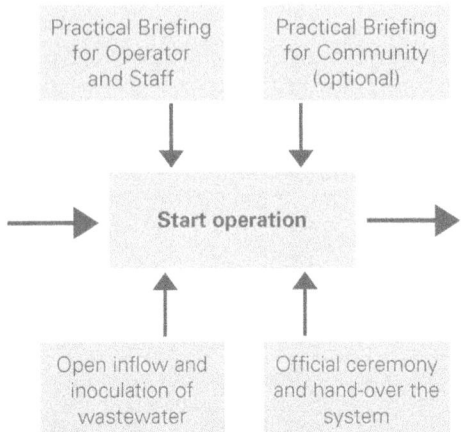

Picture 6_20: Start operation

Picture 6_21: The opening ceremony is welcomed by the residents

6.6.2 Operation & maintenance

Operation and maintenance should be backed by a detailed contract. Depending on the approach, the community operating body or an external operator, such as an NGO, a public entity or private company, can be responsible for the operation and maintenance. The lead agency should sign formal agreements with the relevant parties, clearly defining the terms of reference. A contract with a service provider could, for example, include the following terms:

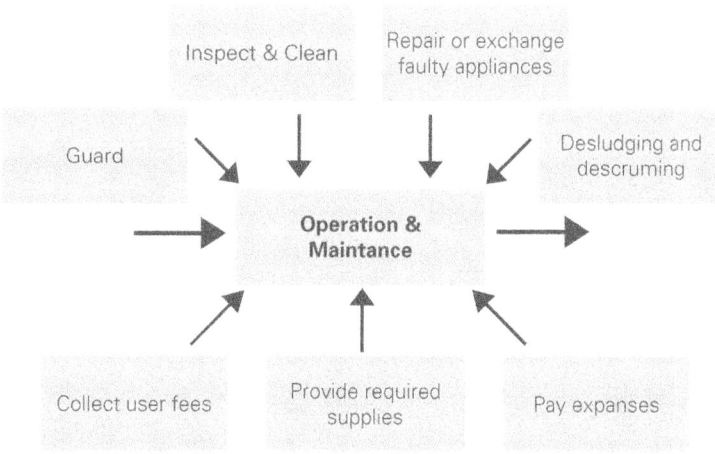

Figure 6_22:
Possible operation and maintenance tasks

- community sanitation centres have to be operated and guarded from 5:00 until 22:00
- the entrance areas (terraces) must always be maintained properly (must be cleaned at least twice a day)
- toilets, shower cells, laundry places and the rest of the plot must be inspected and cleaned daily
- faulty appliances, such as light bulbs and leaking pipes, must be replaced.
- the seal of the bio-digester has to be checked for gas tightness and water has to be re-filled on a weekly basis
- inspection chambers have to be checked and cleaned every week

- the water tank has to be cleaned and cobwebs have to be removed every month.
- every six month, the system must be inspected by professional technical staff, who will sample wastewater, analyse of effluent water and de-scum treatment modules.
- the treatment system must be desludged every two years
- user fees must be collected
- all operation and maintenance activities must be documented

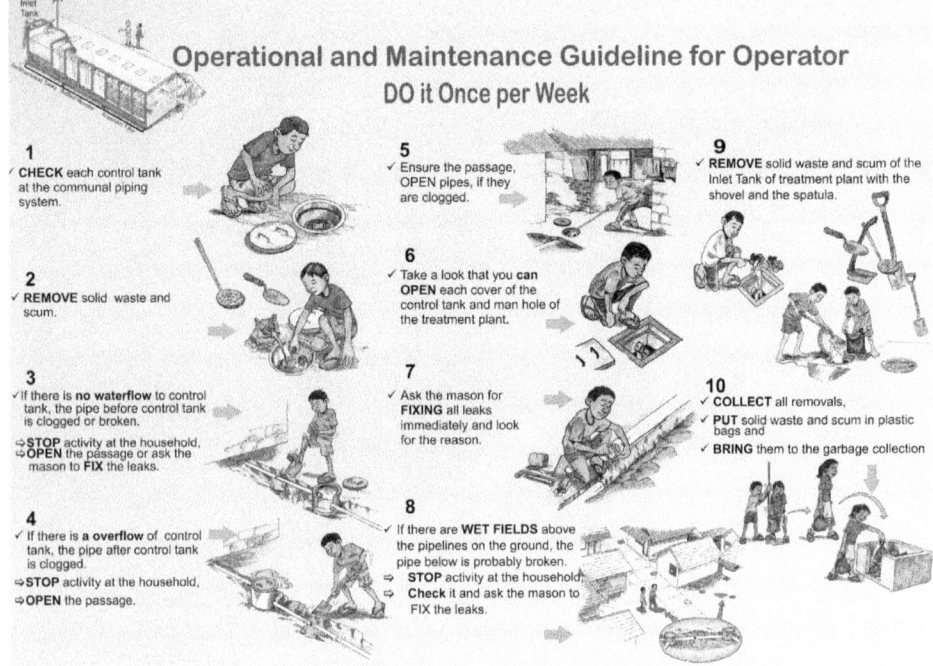

Picture 6_23: Operation & maintenance manual

As described in section 11.3,, sludge collection, treatment and disposal are an integral part of an overall wastewater-management scheme. On-site sludge treatment and disposal is usually not advisable in urban and suburban areas. In such cases, infrastructure must be provided to allow environmentally sound off-site handling.
Some municipalities have a sludge-treatment facility, which can be used. Where none is available, the responsibiling for establishing central sludge-treatment and disposal facility usually lies with the local government or municipality.

To permit the use of vacuum trucks, treatment facilities, which require regular desludging, should not be more than 50m (length of flexible desludging pipes) away from a street accessible by such a vehicle. The truck's sludge container is connected to a sludge pump with a flexible pipe. The pipe is put through the inspection shaft to the bottom of the treatment tank before the vacuum pump is switched on. During desludging only matured "black" sludge should be removed. Establishing this type of sludge-management service is one of the main challenges for public bodies. Without such services wastewater-treatment systems stop functioning. Without adequate regulations and law enforcement many houholds and public and private entities practise uncontrolled discharge of their sludge, leading to great environmental risks and health hazards.

ure 6_24:
Monitoring is required to ensure that service providers handle, treat and dispose sludge in accordance with regulations

ure 6_25:
Operation & maintenance manual for DEWATS developed by the indian partner CDD download: www.borda-sa.org/uploads/o&m_manual-lowres.pdf)

Even if the desludging infrastructure has been established, the services must be monitored. Positive results have been achieved with a model, in which users pay the service provider with a "chip", for which the service provider gets reimbursed by the municipality after delivering the sludge to the treatment facility; users purchase the "chips" from representatives of the local government. This system prevents corruption and the service provider dumping the waste.

6.6.3 Use of biogas

Details on biogas utilisation are given in section 11.5 (page 325). This section focusses on the use of biogas with regard to the operation phase of CBS programmes.

The use of DEWATS-generated biogas is highly recommended because it:
- makes use of a renewable-energy source
- reduces greenhouse gases, which would otherwise escape from the treatment process
Further details are given in section 4.4.2 (page 77).

The efficient and sustainable use of the biogas requires:
- sufficient production of biogas (see sections 10.2.4, page 241 and 11.5, page 325)
- maintenance of the biogas equipment (see also section 11.5, page 325)
- clear definition of who is entitled to use the biogas

Within CBS projects, biogas is normally used for applications, such as cooking, water heating or lighting. Experience shows that tensions can arise within the community if it is unclear who is entitled to use the biogas – leading to a waste of the resource in some projects. Positive results were achieved where the service providers or individuals responsible for the operation and maintenance of the overall system received the benefit. Since they handle the CBS on a daily basis, these individuals already have a deeper understanding of biogas production and can be trained to incorporate the maintenance of the biogas equipment into their other maintenance duties.

Maintenance duties for biogas appliances include:
- cleaning biogas burners and pipes to prevent clogging with water vapours
- replacing biogas lamp mantles regulary

The project leaders should make sure the decision about biogas use is made early on in the project. Once the stakeholders agree, the future users can be trained to maintain the biogas equipment, while it is being installed.

Picture 6_26: Only fully functioning systems are accepted

Picture 6_27: Biogas equipment has to be well-maintained

6.6.4 Monitoring and evaluation

The performance of the wastewater-treatment system should be checked every 6 to 12 months. The inlet and outlet quality should be analysed to verify that legal standards are being met. Results can be compared to the target performance (planning phase) to optimise the design of future plants.

Daily records should be kept by the operator and operating body so that the service can be evaluated. Records should include data on the daily number of users, specific problems and the operation and maintenance activities carried out. Where systems are badly operated, or the number of users decreases with time; the local authority should investigate the reasons and take appropriate action. This might include replacing the service providers or revising the operation and maintenance scheme.

Picture 6_28:
Monitoring of the operation and maintenance scheme contributes to sound treatment performance of CBS solutions

DEWATS components & design principles

DEWATS can be constructed and operated successfully almost anywhere because they rely on natural wastewater-treatment processes, without special equipment, chemicals, or energy supply. This chapter explains the treatment processes and how they apply to different DEWATS components, in order to guide the reader in appropriate technical selection and design.

The chapter is sub-divided into the following sections:
- basics of wastewater treatment
- parameters for wastewater-treatment design
- DEWATS – technical components
- dimensioning of DEWATS

7.1 Basics of wastewater treatment

7.1.1 Definitions: pollution & treatment

Pollution is the undesirable state of the environment being contaminated with substances, which disturb the natural balance of nature and can lead to health consequences for flora, fauna and humans.

Although domestic wastewater is mainly organic, the high concentration of the substances has a polluting effect on open-water bodies, groundwater or soil, due to the oxygen-draining chemical and bio-chemical reactions that result.

Pathogens, including helminth eggs, protozoal cysts, bacteria and viruses, are responsible for innumerable cases of disease and death in the world.

Phosphorus and nitrogen are essential nutrients for plant growth. Their introduction to water bodies can generate great algae populations, which limit the amount of sunlight that can shine into the water, thereby leading to excessive oxygen consumption within the water body until other aquatic life-forms can no longer survive. Furthermore, nitrogen is poisonous to fish in the form of ammonia gases and may also become poisonous to other life-forms, including humans, in the form of nitrite.

Most heavy metals are toxic or carcinogenic. They harm the aquatic life of the receiving water and affect humans through the food chain.

Treatment consists of a wide range of procedures that relieve the negative effect of the pollutants, by removing or changing harmful substances into a harmless or less-harmful state. DEWATS treatment depends on natural bio-chemical and physical processes including:
- degradation of organic matter until the point at which chemical or biological reactions stop (stabilisation)
- physical separation and removal of solids from liquids
- removal or transformation of toxic or otherwise-dangerous substances (for example, heavy metals or phosphorous), which are likely to distort sustainable biological cycles, even after stabilisation of the organic matter

7.1.2 Biological treatment

Stabilisation occurs through degradation of organic substances via chemical processes, which are biologically mediated (bio-chemical processes). The processes are the result of the metabolism by micro-organisms, in which complex and high-energy molecules are transformed into simpler, low-energy molecules. Metabolism is the break-down of organic matter (from feed to faeces) to gain energy for life, in this case for the life of micro-organisms, which store and release the gained energy in the form of ATP (adenosine triphosphate). A few chemical reactions happen without the help of micro-organisms. Most of the micro-organisms involved are biologically classified either as bacteria or as archaea. In the past, archae were viewed as an unusual group of bacteria (archaebacteria). Due to their different evolutionary history, they are now classified as a separate domain. That is why "methanobacteria" according current classification are no longer bacteria but archae. In order to avoid confusion, the generic term "micro-organisms" is used.
In the main, wastewater treatment is the degradation of organic compounds, and subsequent oxidisation of carbon (C) to carbon dioxide (CO_2), nitrogen (N) to nitrate (NO_3), phosphorus (P) to phosphate (PO_4) and sulphur (S) to sulphate (SO_4). Hydrogen (H) is also oxidised to water (H_2O). In anaerobic processes, some of the sulphur is formed into hydrogen sulphide (H_2S), producing the typical "rotten-egg smell". The largest amount of oxygen (O) is required for burning carbon ("wet combustion").

The process of oxidation happens aerobically with free dissolved oxygen (DO) present in water, or anaerobically without oxygen from outside the degrading molecules. Anoxic oxidation takes place when oxygen is taken from other organic substances such as nitrate or sulphate.

Facultative processes include aerobic, anoxic and anaerobic conditions, which prevail at the same time at various parts of the same vessel or at the same place after each other. In anoxic respiration and anaerobic fermentation, as there is no free oxygen available, all oxygen must come from within the substrates. Anaerobic treatment is never as complete as aerobic treatment because there is not enough oxygen available within the substrate itself. The chemical reactions under aerobic, anoxic and anaerobic conditions are illustrated by the decomposition of glucose:

Decomposition via aerobic respiration: $C_6H_{12}O_6 + 8O_2 = CO_2 + 6H_2O$

Decomposition via anoxic respiration: $C_6H_{12}O_6 + 4NO_3 = 6CO_2 + 6H_2O + 2N_2$

Decomposition via anaerobic fermentation: $C_6H_{12}O_6 = 3CH_4 + 3CO_2$

Micro-organisms need nutrients to grow. Any living cell consists of C, H, O, N, P and S atoms. Consequently, any biological degradation demands N, P and S atoms beside C, H and O. Trace elements are also needed to form specific enzymes. Enzymes are specialised molecules, which act as a kind of "key" to "open-up" complex molecules for further degradation.

Carbohydrates and fats (lipids) are composed of C, O and H atoms and cannot be fermented in pure form (Lipids are "ester" of alcohol and fatty acids; an ester is a composition that occurs when water separates off). Proteins are composed of several amino acids. Each amino acid is composed of a COOH-group and a NH_3-group plus P, S, Mg or other necessary trace elements. Thus, proteins contain all the necessary elements and, consequently, can be fermented alone. A favourable proportion between C, N, P and S (varying around a range of 50:4:1:1) is a pre-condition for optimum treatment.

7.1.3 Aerobic – anaerobic

Aerobic decomposition takes place when dissolved oxygen is present in water. Composting is also an aerobic process. Anoxic digestion occurs when dissolved oxygen is not available, but bacteria get oxygen for energy "combustion" by breaking it away from other, mostly organic substances present in wastewater, predominantly from nitric oxides. Anaerobic digestion breaks up molecules composed of oxygen and carbon to ferment them to carbohydrates.

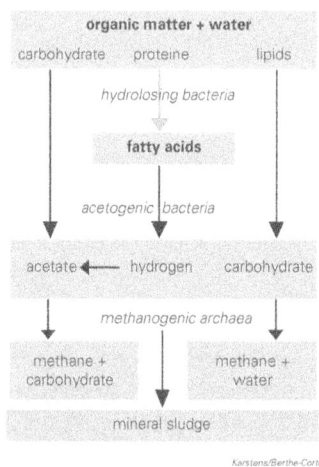
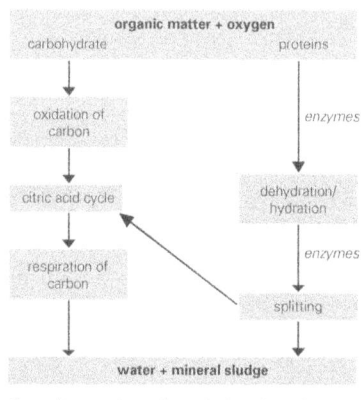

Picture 7_1: The anaerobic process in princ[iple]

Picture 7_2: The aerobic process in princ[iple]

The aerobic process happens much faster than anaerobic digestion and therefore dominates when free oxygen is available. The high rate at which decomposition takes place is caused by the shorter reproduction cycles of aerobic bacteria as compared to anaerobic micro-organisms. The latter leave some of the energy unused, which is released in the form of biogas. Aerobic micro-organisms use a larger portion of the pollution load (about 50% of the COD) for production of their own bacterial mass compared to anaerobic ones (only about 5% of the COD). That is why anaerobic processes produce 90% less sludge compared with aerobic ones. For the same reason, anaerobic sludge is less slimy than aerobic sludge and is easier to drain and dry.

Aerobic treatment is highly efficient when there is enough oxygen available. However, compact aerobic treatment tanks need external oxygen, which must artificially be supplied by blowing or via surface agitation. Such technical input consumes technical energy.

The anaerobic treatment process proceeds at a lower rate. It benefits from a higher digestion temperature. Therefore, it is well suitable for DEWATS in tropical and subtropical countries. Ambient temperatures between 15° and 40°C are sufficient. Anaerobic digestion (fermentation) releases biogas (CH_4 + CO_2), which can be used as a fuel (see section 6.6.3, page 129).

7.1.4 Physical treatment processes

Wastewater treatment relies on the separation of solids, both before and after stabilisation. Even dissolved particles are decomposed into the three main fractions: water, gases and solids, of which the solids have to be removed. The choice of method for solid removal depends on the size and specific weight of the suspended solids.

Screening

Screening of larger solids is the foremost step in conventional treatment plants. In DEWATS, screening is not advisable because screens require cleaning at very short intervals, i.e. daily or weekly, which demands a safe storage and treatment space in the immediate vicinity for the removed screenings. A blocked screen is an obstacle that plugs the entrance of the plant. DEWATS should allow for the full amount of wastewater to pass through the plant without obstructions. If this fails, it may happen – and, in fact, happens quite often – that the operator "organises" a trouble-free by-pass, which pollutes the environment, as if the treatment plant did not exist. For this reason it is recommended to avoid screens and, instead, provide sufficient additional space to accommodate larger solids within the first sedimentation chamber.

Sedimentation

Separation of solids happens primarily by gravity, predominantly through sedimentation. Coarse and heavy particles settle within a few minutes or hours, while smaller and lighter particles may need days and weeks to finally sink to the bottom. Small particles may cling together, forming larger flocs that also sink quickly. Such flocculation happens when there is enough time and little to no turbulence; stirring hinders quick sedimentation. Sedimentation is slow in highly viscose substrate.

Sedimentation of sand and other discrete particles works best in vessels with a relatively large area. These vessels may be shallow, since depths of more than 50cm have no influence on the sedimentation process in the case of discrete particles.

grain size in mm	1	0.5	0.2	0.1	0.05	0.01	0.005
quartz sand	502	258	82	24	6.1	0.3	0.06
coal	152	76	26	7.6	1.5	0.08	0.015
SS in domestic wastewater	120	60	15	3	0.75	0.03	0.008

Table 10:
Settling velocities of coarse particles (m/h).

Suspended sludge particles have settling properties different from coarse particles.
Source: K.+K. Imhe p. 126

This is different for finer coagulant particles, where sedimentation increases with basin depth. This is because settling particles meet suspended particles to form flocs which continue to grow larger and settle faster on their way to the bottom. A slow and non-turbulent flow – still and undisturbed water – supports "natural" coagulation for sedimentation.

DEWATS components & design principles

Settled particles accumulate at the bottom. In the case of wastewater, any sediment also contains organic substances, which begin to decompose. This decomposition, which occurs in any sludge sedimentation basin and, to a lesser extent, in grit chambers, results in the formation of carbon dioxide, methane and other gases. These gases are trapped in sludge particles of the vessel which then float to the top when the numbers of gas molecules increase. This process not only causes turbulence; it also ruins the success of the sedimentation that has taken place. The Imhoff tank, through its baffles, prevents such gas-driven particles from surging and spoiling the effluent. The UASB process deliberately utilises this balance of sedimentation (= downstream velocity) and up-flow of sludge particles (= upstream velocity).

After decomposition and the release of gases, the stabilised (mineralised) sludge settles permanently at the bottom, where it accumulates and occupies tank volume. It must be removed at regular intervals. Since many pathogens, especially helminths, also settle well, sedimentation plays an important role in the hygienic safety of domestic or husbandry wastewater treatment.

Flotation

Flotation is the predominant method for fat, grease and oil removal. In conventional wastewater treatment the process is also used to remove small particles by injecting fine air bubbles to the bottom of the tank.

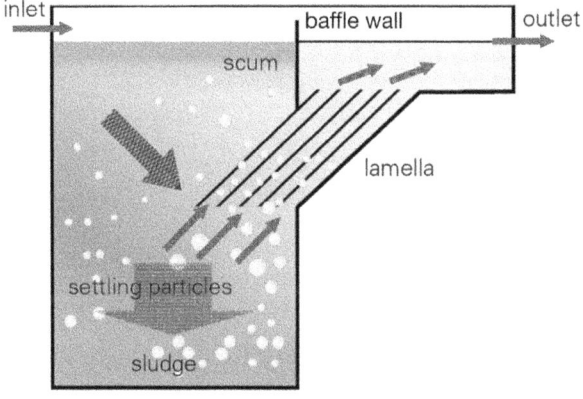

Figure 7_3:
Principle of lamella solids separator to improve sedimentation – lamella may be made of plastic sheets, concrete slabs or PVC pipes

Most fatty matter can be identified by simple observation tests, similar to settleable solids. If fats, which are detected by laboratory analysis, are not separated by floatation, they present themselves as colloids, which can only be removed after pre-treatment (e.g. after acidification).

Unwanted flotation occurs in septic tanks and other anaerobic systems, where floating layers of scum may form. Accumulating scum can be removed manually, or can be left purposely to "seal" the surface of anaerobic ponds, preventing bad odour.

Flotation and sedimentation can be improved by installing slanted lamella sheets or several layers of slanting pipes. These surfaces artificially increase the separation of solids from liquids by facilitating floc and gas accumulation.

Picture 7_4:
Baffle wall retaining scum, inlet is at the right side, water flows underneath the downflow baffle into the compartment at the left side

Filtration

Filtration is necessary for the removal of suspended solids, which do not "self-flocculate", settle or float within a reasonable time. Most filters have a double function: while forming a physical obstacle for smaller solid particles, they also provide a fixed surface on which treatment micro-organisms can grow. Both organic growth of micro-organisms and accumulated solids can lead to the clogging of the filter. Physical filters retain solids which accumulate, unless they are removed. Coarse filters, where physical filtration occurs primarily with the help of micro-organisms, can be cleaned by back-flushing. In this way, micro-organisms and suspended solids are flushed away simultaneously as, for example, is typically done with trickling filters. Upstream filters may be back-flushed. The filter media of sand and finer gravel filters must be removed, cleaned and replaced after several years of use.

Needless to say, filters with smaller grain size provide more efficient particle removal. On the other hand, effective filtration requires the retention of many solids and, therefore, leads to faster clogging. The permeability and durability of filters is always reciprocal to its treatment efficiency. Filter material of round and almost equal grain size is more efficient and renders longer service than filters of mixed grain sizes.

Aerobic filters produce more sludge than anaerobic filters and, consequently, block faster. However, they also have a self-cleaning effect when given sufficient resting time, as the aerobic bacteria in the sludge practise a kind of "cannibalism" (autolysis) when the nutrient supply stops.

Sludge accumulation

Sedimentation and filtration lead to sludge accumulation at the bottom of vessels. With time, the sludge compacts; consequently, older sludge occupies less volume than fresh sludge. Sludge-removal intervals are, therefore, important design criteria. The sludge must be handled and treated adaquately (see section 11.3).

7.1.5 Elimination of pollutants

Elimination of nitrogen

Nitrogen removal occurs in two steps: nitrification followed by denitrification, which results in pure nitrogen diffusing into the atmosphere.

Nitrification is oxidation. Nitrate is the most stable form of nitrogen and its presence indicates complete oxidation. Denitrification is reduction, or the separation of that very oxygen from the oxidised nitrogen. The pure gaseous nitrogen that remains is insoluble in water and, therefore, evaporates easily. Nitrogen escaping from the denitrification process may cause floating foam or scum, similar to the effect seen from the gas release by settled anaerobic sludge. Since nitrogen is the major compound of air it is ecologically harmless.

During nitrification NH_3 (ammonia) is oxidised by two special groups of bacteria – nitrosoma convert ammonia to nitrite (NO_2) and nitrobacter convert nitrite to NO_3 (nitrate). Since nitrobacter grow slowly, a higher sludge age and, thereby, longer retention time is needed for oxidation of nitrogen (= nitrification) than is required for oxidation of carbon (see section 8.1.1 "Control parameters", page 159).

Denitrification occurs faster than nitrification, as several groups of bacteria are able to utilise nitrate oxygen under anoxic conditions (absence of free oxygen). Incomplete denitrification may lead to formation of the poisonous nitrite (NO_2), instead of nitrate (NO_3).

This happens because the time left for the bacteria to consume all the oxygen is not enough or because there is not enough organic material left to absorb the NO_3-oxygen. Some non-DEWATS treatment processes recycle nutritious sludge to prevent such nutrient deficiency. A certain amount of nitrate in the effluent could also be a source of oxygen for the receiving water. In DEWATS, nitrate removal usually does not receive special attention, in that additional technical measures are not taken.

Elimination of phosphorus

Micro-organisms cannot transform phosphorus into a form in which it loses its fertiliser quality permanently. Phosphorus compounds remain potential phosphate suppliers. This implies that no appropriate biological process, either aerobic or anaerobic can remove phosphorus from wastewater. Phosphorus removal from water "normally" takes place by removal of bacteria mass (active sludge) or by removal of phosphate fixing solids via sedimentation or flocculation. Iron chloride, aluminium sulphate or lime fixes phosphates, a fact that can be utilised by selecting suitable soils in ground filters. However, the removal of phosphorus in root zone filters has not proven to be as efficient and sustainable as expected by the pioneers of these systems.

Elimination of toxic substances

Heavy-metal compounds occuring in bigger molecular structures may settle easily. Their removal is not difficult. Heavy-metal contaminated sludge must be handled accordingly and disposed of safely, at proper landfill sites. Heavy-metals occuring in the form of dissolved ions do not settle at all. Along with other soluble toxic substances, they are difficult to remove. There are numerous ways of eliminating or transforming toxins into non-toxic matter, which cannot be described here. You should consult more specialised literature.

noxious substance group	1 NSU is eqal to
oxidisable matter	50kgCOD
phosphorous	3kgP
nitrogen	25kgN
organic fixed halogenes	2kgAOX
mercury	20gHg
cadmium	100gCd
chromium	500gCr
nickel	500gNi
lead	500gPb
copper	1,000gCu
dilution factor for fish toxicity	3,000m^3

Table 11:
Noxious Substances Units (NSU) according to German federal wastwater charges act
Mercury is the most dangerous substance on the list.
Source: Imhoff, 1990

High salt content inhibits biological treatment and is very difficult to remove. In the case of saline water used for domestic or industrial purposes, for example, the water remains saline even after treatment. So it should not be used for irrigation and should not be allowed to enter the groundwater table or receiving rivers that carry too little water.

Removal of pathogens

Even after treatment, wastewater should be handled carefully. Underground filtration and large pond systems are relatively efficient in pathogen removal, but not necessarily to the extent that wastewater can be called safe for bathing – let alone drinking. However, reuse for irrigation is safe under certain conditions (see section 11.4).

Helminth eggs and protozoa accumulate in sediment sludge, so are largely retained inside the treatment system, where they stay alive for several weeks. Most micro-organisms and viruses bound to the sludge die more quickly. Pathogens, which are not caught in the sludge and remain suspended in the effluent, are hardly affected. This is especially true in high-rate reactors, like filters or activated-sludge tanks. These bacteria and viruses exit the plant alive, although the risk of virus infection from wastewater has proven to be low.

type of infection, type of wastewater	country	dose of chlorine g/m³	contact time h	total rest chloride mg/l
intestinal pathogens	China		> 1.0	5
tubercular pathogens	China		> 1.5	7
raw wastewater	Germany	10 - 30	0.25	traces
post treatment	India	3		
post treatment	Germany	2	0.25	traces
odour control	Germany	4	0.25	traces

Table 12:
Comparing the use of chlorine for different requirements various places.
Different sources

DEWATS components & design principles

Exposure to UV rays has a substantial hygienic effect, in addition to sedimentation, predation and die-off in a hostile environment. The highest rate of pathogen removal can be expected from shallow ponds with long retention times, for example three ponds in series with HRT of 8 to 10 days each. Constructed wetlands with their multifunctional bacterial life in the root zones can also be very effective. However, it is the handling after treatment, that ensures hygienic standards.

Using chlorination to kill pathogens in wastewater is only advisable for hospitals in the case of epidemics and similar circumstances. It may also be applied in slaughterhouse treatment plants, which are only a short distance from a domestic water source. Permanent chlorination is never advisable, as it has adverse effects on the environment: Water is made unsuitable for aquatic life, and chlorine can react with organic matter to produce dangerous chemicals.

Bleaching powder (chlorinated lime) containing approximately 25% free chlorine is the most common source of chlorine. Granular HTH (high test hyperchlorite) containing 60 to 70% Cl is available on the market under different brand names. Since chlorination should not be a permanent practice, a chamber for batch supply, followed by a contact tank of 0.5 – 1h HRT will be sufficient (picture 7_5).

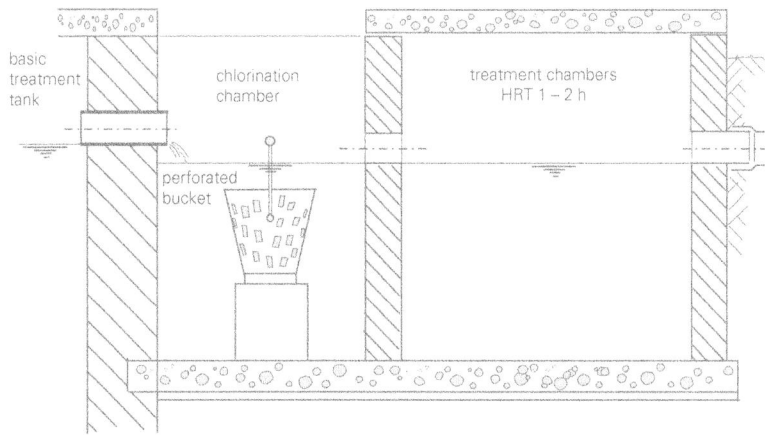

Picture 7_5: Post-treatment chlorination carried out in a batch chamber for small scale applications. The bucket is filled with bleaching powder, which is dissolves automatically. This plant is acceptable for emergency disinfection of effluent from rural hospitals only because controlled dosing is not possible.

7.1.6 Ecology and self-purification in nature

An understanding of the self-purification ability of the natural environment helps in designing DEWATS intelligently. On the one hand, only harmless wastewater should be discharged; on the other hand, nature may be incorporated into the design for the completion of the treatment processes.

Surface water

The biological self-purification effect of surface waters depends on the climate, weather and on the relative pollution load in the water. The presence of free oxygen is a precondition for the self-purification process. The higher the temperature, the higher the rate at which the degrading bacteria, which are responsible for purification, multiply. At the same time, the intake of oxygen via surface increases, but oxygen solubility drops with increasing temperature. Rain and wind increase the oxygen-intake capacity. Consequently, acceptable pollution loads or wastewater volumes must be dimensioned according to the season with the least favourable conditions
(for example, winter or summer in temperate zones, dry season in the tropics).
It is difficult to reanimate water once the self-purification effect has stopped as from then on, it enters the anaerobic stage.
Extreme seasonal changes make it difficult to maintain the self-purification effect of water throughout the year. However, nature has a way of helping itself, as in the case of lakes and rivers that dry out in long dry seasons when the remains of organic matter compost and are fully mineralised before the next rains come.

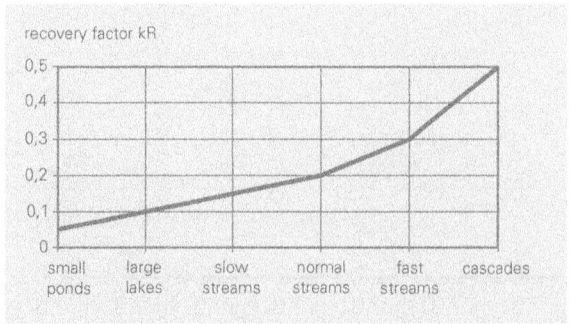

Picture 7_6:
Ability of surface waters to recover oxygen after pollution. Turbulence increases oxygen intake and reduces time for recovery
(Garg, p. 220)

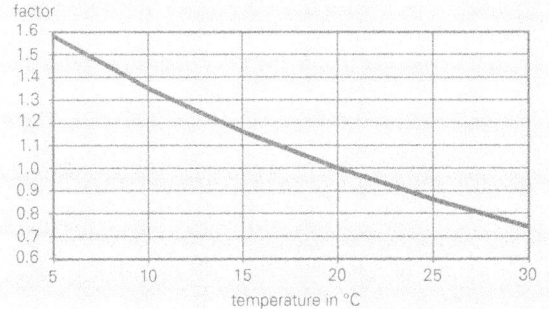

Figure 7_7:
Oxygen intake of natural waters via surface contact. Relative concentration of dissolved oxygen decreases with increasing temperature.

Minerals retain their fertilising quality even after drying. This is why it is better to bring sludge at the bottom of dried lakes, canals or rivers to the fields before it is washed away into the receiving water by the first heavy rains and its rich nutrient value is lost. However, the content of toxic matter in sludge should be observed.

The most important source of oxygen for natural water in an ecosystem is oxygen from the air, which dissolves in water via surface contact. Floating fat, grease or oil films restrict oxygen transmission from the air and, moreover, require additional oxygen for their decomposition.

The nutrients contained in wastewater increase algae growth. In a healthy ecosystem, algae produce oxygen during the day and consume part of this oxygen at night. If the algae population were to become unduly dense, sunlight would not be able to penetrate the dark-green water. As a result, the algae would consume oxygen during the day as well – and the supply of free oxygen that is needed for aquatic life would decrease.

The degree of pollution and, particular, in the content of dissolved oxygen (DO), can be gauged by the variety of plant and animal species found in the water. The colour of the water of rivers and lakes is yet another indicator of the quality of the water. Green or green-brownish water is indicative of high nutrient supply due to algae; a reddish-rosy colour indicates facultative algae and a severe lack of free oxygen; black is often indicative of complete anaerobic conditions of suspended matter.

Nitrogen in the form of nitrate (NO_3) is the main polluting nutrient. In the form of ammonia (NH_3, toxic to fish) it is also a major, oxygen-consuming toxic substance, therfore nitrogen should be kept away from living waters; notwithstanding that nitrate may also function as an oxygen donor in certain instances.

The next most-important polluting nutrient is phosphorus, which is mainly present in the form of hydrogen phosphate (H_2PO_4). Since phosphorus is often the limiting factor for the utilisation of other nutrients, its presence in surface waters is dangerous, as even in small doses it may lead to an oversupply of nutrients (eutrophication). Nitrogen that is normally plentiful needs 10% of phosphorus to be assimilated by plants. That means phosphorus activates ten times as much nitrogen and, therefore may be considered the most polluting element to any receiving water. At the same time, it is this property that makes wastewater rich in phosphate an excellent fertiliser when used for irrigation in agriculture.

Picture 7_8:
DEWATS at a hotel in Cochin, India; large-size ponds are used for post-treatment

Phosphorus accumulates in closed ecosystems, for example in lakes. Unlike nitrogen that can be eliminated as N_2 or N_2O_2 etc., phosphorus remains potentially active in the residue of dead plants, which have previously incorporated it.
For example, phosphate fixed by iron salts can be set free under anaerobic conditions in the bottom sludge, where it is available for new plant growth. It is for this reason that continuous supply of phosphate into lakes is prohibited. While this may seem less dangerous for flowing waters, it must be realised that all rivers end somewhere, at which point phosphorus will accumulate.

While chlorine may be used for disinfecting effluents from hospitals and slaughter-houses, it must be remembered that chlorine also disinfects the receiving waters, thereby reducing their self-purification ability. Moreover, they form chlorinated compounds being potentially carcinogenic.

It is self-evident that toxic substances should not enter any living water. Most toxic substances become harmless in the short term, particularly if they are sufficiently diluted. However, most toxic materials are taken in by plants and living creatures and, in the long run, accumulate in the aquatic lifecycle. Fish from such waters become unsuitable for human consumption and heavy metals accumulate in the bottom sludge of receiving waters, where they remain as a time-bomb for the future.

Groundwater

Groundwater was once rainwater. It is the most important source of water for domestic use, irrigation and other purposes. The supply of groundwater is not infinite. To be sustainable, it must be recharged. Rather than simply draining used water into rivers that carry it to the sea, it would be better to purify this water and use it to recharge the groundwater.

Organic pollution of groundwater happens in cases where wastewater enters underground water-streams directly. A crack-free, 3m-thick soil layer above groundwater is sufficient to prevent organic pollution. Pollution by mineralised matters is possible, however, as salts like nitrate and phosphate are soluble in water and cannot be eliminated by physical filtration when passing through soil or sand layers. Some pathogens may also reach the groundwater despite soil filtration. Viruses can be dangerous, due to their infectious potential, irrespective of their absolute number.

Nitrate is readily soluble in water. So it is easily leached out from soil into groundwater, especially in sandy soil during periods when vegetation is low (for example, winter in cold climates). Groundwater, therefore, will always contain a certain amount of nitrate (mostly above 10mg/l).

Nitrate (NO_3) in itself is rather harmless. For example, in the European Union, drinkingwater may legally contain nitrate up to 25mg/l. It is, however, latently dangerous, as nitrate is capable of changing to nitrite (NO_2) under certain biological or chemical circumstances. This process can even occur inside human blood, where nitrite attaches itself to haemoglobin, reducing the capacity of the haemoglobin to "transport" oxygen – leading to suffocation. Nitrite poses the greatest risk to babies, who have a greater tendency to form nitrite. For this reason, water used for the production of baby food must always contain less than 10mg/l NO_3.

Soil

Pollution can render soils useless for agricultural production. For example, the pH may drop as a result of incomplete anaerobic digestion of organic matter. This is particularly common in clay or loamy soils, where oxygen supply is insufficient due to the physical closure of pores in the soil by suspended solids from wastewater irrigation. Furthermore, soil pollution poses a threat because of washout effects that harm surface- and groundwater alike. Mineral salts in small doses do not pose a problem for wastewater treatment. Using saline wastewater for irrigation over a long period of time, however, may cause complete and irreversible salination of the topsoil. Clay and loamy soils with slow downward percolation are the most affected, as water evaporates from the top layers, leaving the salt behind.

On the other hand, sandy soils may benefit from irrigation with wastewater even when the organic load is high, provided that oxygen can be supplied to deeper soil layers. Well-treated wastewater, containing mineralised nitrogen, phosphorus and other trace elements, can improve soil conditions and is environmentally safe, as long as the application of nutrients is balanced with its in-take by plants. Applying of treated wastewater throughout the year, regardless of demand, may have adverse effects. Nutrients will be leached out into water bodies at times when plant growth is negligible, with the result that nutrients are not available to the plants when needed.

Treatment in DEWATS

DEWATS make use of the natural biological- and physical-treatment processes discussed above to reduce and remove pollutants from wastewater. External energy supply, dosing of chemicals and movable parts are avoided to minimise both possible flaws in operation and maintenance.

As the various natural-treatment processes require different boundary conditions to function efficiently, DEWATS are comprised of a series of treatment units, each providing an ideal environment for the removal of certain groups of pollutants. Stability of the treatment system is ensured, as each treatment step only removes the "easy part" of the pollution load, sending the leftovers to the following step.

| Sedimentation | → | Anaerobic digestion | → | Aerobic decomposition | → | Post-sedimentation |
| removal of easily settleable solids | | removal of easily degradeable organic solids | | removal of more difficult degradable solids | | removal of digested solids and active bacteria mass |

Picture 8_1:
Several steps are required for full treatment

The term "phase separation" has a double meaning. On the one hand it is used for the separation of gas, liquid and solids in anaerobic reactors; on the other hand it is used to describe the technical separation of different stages of the treatment process, either in different locations or in sequences of time intervals. The latter kind of phase separation becomes necessary when suitable nutrients cannot be provided simultaneously to micro-organisms, which have differing growth rates and prefer different feeds. Some micro-organisms grow at a slower rate than others. As not all the enzymes required for degradation are initially found in all substrates, the micro-organisms take time to produce adequate amounts of the missing enzymes. As disscussed previously, enzymes act as the "key which opens the lock of the food box for micro-organisms".

Substrates, for which enzymes are immediately available, can be readily degraded; substrates, which first require the microbial production of specific enzymes, are degraded much more slowly. In an environment which hosts substances that are both easy and difficult to degrade, the microbial population responsible for easy degradation tends to predominate.

To protect the "weaker" (slower) micro-organisms, it is advisable to artificially separate microbial populations in phases by providing each with its own favourable environment. The characteristics of the wastewater and the desired treatment results must be identified, before the dimensions of the treatment vessels for the different phases can be designed.

In the case of DEWATS, it is often easiest to provide longer retention times, so that the "slow" micro-organisms find their food after the "fast" ones have satisfied their demand. This process is easier to manage and, in the case of smaller plants, it is cheaper to design certain units this way. In other units, like the baffled reactor, the efficiency of the treatment in subsequent chambers justifies its higher cost; processes, which require sequencing batch operation involving technical equipment and process control, are thereby avoided.

Phase separation becomes unavoidable if different phases require either anaerobic or aerobic conditions. In the case of nitrogen removal, longer retention times alone do not provide adequate treatment conditions because the nitrifying phase needs an aerobic environment, while denitrification requires an anoxic environment. Anoxic means that nitrate (NO_3) oxygen is available, but free oxygen is not. Anaerobic means that neither free oxygen nor nitrate-oxygen is available. Nevertheless, the aerobic phase can only lead to nitrification if the retention time is long enough for the "slow" nitrifying bacterium to act, as compared to the "fast" carbon oxidisers.

In the case of the addition of plant material to an anaerobic digester, pre-composting of plant residues before anaerobic digestion is another example of simple phase separation. As lignin cannot be digested anaerobically (it requires peroxidase enzymes usually produced by fungi), it is decomposed aerobically. Afterwards, anaerobic micro-organisms can reach the inner parts of the plant material in the digester.

8.1 Parameters for wastewater-treatment design

Treatment must remove or reduce pollutants within the wastewater sufficiently to prevent harm to the environment and humans. Before deciding, what kind of treatment is necessary and the dimensions of each unit, planners and designers must identify the following:
- quality and quantity of the raw wastewater
- local conditions and their influence on treatment processes
- standards to be fulfilled in final use or discharge

Laboratory analysis is used to determine the quantity and quality of the pollution load, the feasibility of treatment, the environmental impact under local conditions – and whether a particular wastewater is suitable for biogas production. Some parameters can even be seen and understood by experienced observation.

As the quality of wastewater changes according to the time of day and from season to season, the analysis of data is never absolute. It is far more important that the designer understands the significance of each parameter and its "normal" range than to know the exact figures. Ordinarily, an accuracy of ±10% is more than sufficient.

This chapter gives a concise overview, introducing:
- control parameters, essential for characterising wastewater and
- dimensioning parameters, utilised in DEWATS design

Textbooks on the analysis of wastewater should be consulted for laboratory techniques or comprehensive handbooks on wastewater, such as Metcalf and Eddy's *"Wastewater Engineering"*.

8.1.1 Control parameters

Volume

The daily volume or the flow rate of wastewater determines the required size of the building structure – on which the feasibility or suitability of the treatment technology is decided. It is essential not to underestimate the peak flow.

Surprisingly, the determination of flow rate is often rather complicated, due to the fact that flow rates change throughout the day or with the season, and that volumes have to be measured in "full size". It is not possible to take a representative sample. In the case of DEWATS, it is often easier and more practical to measure or enquire about the water consumption (per capita consumption of water from taps and/or wells) rather than try to measure the wastewater production. The flow of wastewater is not directly equal to water consumption, since not all the water that is consumed ends up in the drain (for example, water for gardening), and because wastewater might be a mix of used water and stormwater. If possible, stormwater should be segregated from the treatment system, especially if it is likely to carry substantial amounts of silt or rubbish. Rainwater drains should never be connected to the treatment plant, however, ponds and planted gravel filters will be exposed to rain (and evaporation). The volume of water in itself is normally not a problem as hydraulic loading rates are not likely to be doubled and a certain flushing effect might even be advantageous. Soil clogging (silting) could become a problem, however, if stormwater reaches the planted gravel filter after eroding the surrounding area.

For high-rate reactors, like anaerobic filters, anaerobic baffled reactors and UASB, the flow rate could be a crucial design parameter. If exact flow data are not available, the hours of the day, which account for most of the flow, should be determined and used. Hydraulic retention-time calculations should take into account the flow rate fluctuation.

The flow rate is calculated by collecting and measuring volumes per time period. Possible measurement techniques include monitoring the rise in level of a canal that is closed for a period of time, or the number of buckets filled during a given period. Another good indicator of the actual flow rate is the time it takes, during initial filling for the first tank of a treatment plant to overflow.

In larger plants the flow rates are normally measured with measuring flumes (for example, the Parshall flume) where a rise in level across the flume is related to the flow.

Solids

Total solids (TS) or dry matter (DM) include all matter, which is not water. Organic total solids (OTS) or volatile solids (VS) are the organic fraction of the total solids. TS is found by drying the sample. The inorganic fraction is found by burning the dry matter and weighing the ash. TS minus ash is OTS or VS. Solids may be measured in mg/l or as a percentage of the total volume.

The parameter "suspended solids" (SS) is the amount of organic or inorganic matter that is not dissolved in water. Suspended solids include settleable solids and non-settleable or suspended solids. Settleable solids sink to the bottom within a short time. They can be measured with a standardised procedure in an Imhoff-cone, usually for a defined settling period of 30 minutes, one hour, two hours or one day. Measurement of settleable solids is the easiest method of wastewater analysis because solids are directly visible in any transparent vessel. For collecting initial on-site information, any transparent vessel will do (for example, water bottles, which should be destroyed for hygienic safety after use).

Table 13:
Domestic wastewater derives from various sources. Composition of wastewater depends largely on standard of living and domestic culture.
Source: Metcalf&Eddy, 1996

range source	min g/cap*d	max g/cap*d
faeces (solids, 23%)	32	68
ground food wastes	32	82
wash waters	59	100
toilet (incl. paper)	14	27
urine (solids, 3.7%)	41	68

Non-settleable suspended solids consist of particles which are too small to sink to the bottom within a reasonable time (with regard to the design parameters of the treatment processes). SS is determined by sample filtration. Suspended solids are an important parameter because they cause turbidity in the water and may cause physical clogging of pipes, filters, valves and pumps.

Colloids are very fine suspended solids (< 0.1ηm) which pass laboratory filter paper, but are not dissolved in water (dissolved solids are solutes (molecules or ions) that disperse through out the water molecules). A high proportion of fatty colloids can create large problems in fine-sand filters.

In domestic wastewater, the BOD derives to approximately one third (33%) from settleable solids, to half (50%) from dissolved solids, while one sixth (17%) of the BOD derives from non-settleable SS (see Table 17, page 165).

Fat, grease and oil

Fat and grease are organic matter that is biodegradable. However, since they float on water and have a sticky consistency their physical properties are a problem in the treatment process and in nature. It is best to separate fat and grease before biological treatment and dispose to sludge treatment facilities (see section 11.3) or other specific recycling plants (e.g. soap production).

The amount of fat that remains in treated domestic wastewater is normally small. A fat content of approximately 15 to 60mg/l is allowed in the effluent of slaughterhouses or meat-processing plants for discharge into surface waters. Mineral grease and mineral oils – such as petrol or diesel – although they may also be treated biologically, should be kept away from the treatment system. Their elimination is not within the scope of DEWATS.

Turbidity, colour and odour

Most wastewaters are turbid because the solids suspended in them scatter and disperse light. So highly turbid fluid indicates a high content of suspended solids. Metcalf and Eddy define the relationship between turbidity and suspended solids in the following equation:

$$SS\ [mg/l] = 2.35 \times turbidity\ (NTU),\ or$$

$$turbidity\ (NTU) = SS\ [mg/l] / 2.35^{32}$$

[32] Metcalf & Eddy, 1996; page 257

NTU is the standardised degree of turbidity. Its value can be determined with the help of a turbiditimeter or by standardised methods, which measure the depth of water at which it is no longer possible to see a black cross or a circle against a white background. Turbidity may prevent algae in surface waters from producing oxygen during daytime, as would otherwise be the case.

The colour is not only indicative of the source of wastewater, but also of the state of degradation. Fresh domestic wastewater is grey, while aerobically degraded water tends to be yellow-brown, and water after anaerobic digestion turns blackish. Turbid, black water may be easily settleable because suspended solids sink to the bottom after digestion when given enough undisturbed time to form flocs.
A brownish colour indicates incomplete aerobic or facultative fermentation.

Wastewater that does not smell probably contains enough free oxygen to restrict anaerobic digestion – or the organic matter has long since been degraded. A foul smell ("like rotten eggs") comes from H_2S (hydrogen sulphide), which is produced during anaerobic digestion, especially at a low pH. A foul smell, therefore indicates that free oxygen is not available and that anaerobic digestion is still underway. Vice versa, whenever there is substantial anaerobic digestion there will always be a foul smell.
Fresh wastewaters from various sources have characteristic smells. Experience is the best basis for drawing conclusions: dairy wastewater will smell like dairy wastewater, distillery wastewater will smell like distillery wastewater, etc. "smelling the performance" of treatment plants is an important skill. An alert wastewater engineer should remember the different odours and their causes, to build up a repertoire of experience for future occasions.

COD and BOD

COD (Chemical Oxygen Demand) is the most common parameter for measuring organic pollution. It describes the amount of oxygen required to oxidise all organic and inorganic matter found in the water. The BOD (Biochemical Oxygen Demand) is always smaller than the COD. It describes the amount of oxygen required for the oxidisation of matter, which can be oxidised by the biological organisms in a body of water. It approximates the organic fraction of the COD. Under standardised laboratory conditions at 20°C, it takes about 20 days to activate the total carbonaceous BOD (=$BOD_{ultimate}$, BOD_{total}). In order to save time, BOD-analysis determines the biological oxygen demand after five days. The result is called BOD_5, which in practice, is commonly referred to simply as BOD.

The BOD_5 is a certain fraction (approximately 65 to 70%) of the ultimate BOD. This fraction is different for each wastewater, depending (for example) on the proportions of organic matter in soluble and suspended form. The ratio of BOD_{total} to BOD_5 is larger for refractory or difficult degradable wastewater and, thus, it is also larger for partly treated wastewater.

COD and BOD are the results of standardised laboratory-analysis methods. They do not fully reflect the bio-chemical truth, but are reliable indicators for practical use.

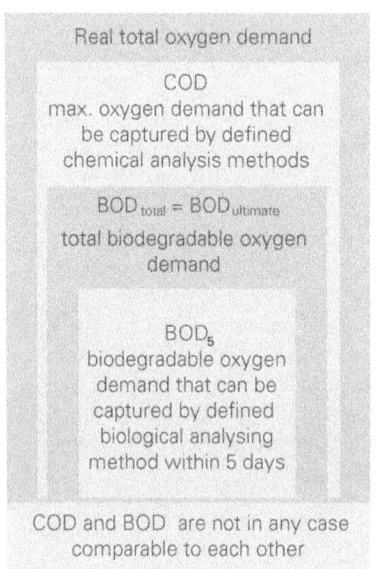

Picture 8_2:
Definition of oxyg demand. The BO is a part of the to BOD; the total BC may be understo as part of the CO and the COD is p of the absolute re oxygen demand. The total BOD ma be equal to the COD; the COD ma be equal to the re oxygen demand

Biological oxygen demand describes the portion of the wastewater which can be digested easily, for example, anaerobically. The COD/BOD_{total} approximates the relation of total oxidisable matter to organic matter, which is degraded first by the most common micro-organisms. For example, if a substrate is toxic to micro-organisms, the BOD is zero; the COD nonetheless may be high, as would be the case with chlorinated water. In general, if the COD is much higher than the BOD (>3 times) one should check the wastewater for toxic or non-biodegradable substances.

In practice, the quickest way to determine toxic substances is to have a look at the shopping list of the institution which produces the wastewater.

The kinds of detergent bought by a hospital may be more revealing than a wastewater sample taken at random. It is important to know that the COD in a laboratory test shows the oxygen donated by the test-substance, which is normally $K_2Cr_2O_7$ (potassium dichromate). The tested substrate is heated to initiate the chemical reaction (combustion). Occasionally, $KMnO_4$ (potassium permanganate) is used for quick on-site tests – also known as the permanganate value (PV) or oxygen adsorbed from permanganate (OA). The COD_{Cr} is approximately twice as much as the COD_{Mn}; however, the two values do not have a fixed relation, which is valid for all wastewaters.

Easily degradable wastewater has a COD/BOD relation of about 2. The COD/BOD ratio widens after biological, especially anaerobic treatment because biological degradation has already taken place. COD and BOD concentrations are measured in mg/l or in g/m^3. Absolute values are measured in g or kg. A weak wastewater from domestic sources, for example, may have a COD below 500mg/l while a strong industrial wastewater may have a value of up to 80,000mg/l BOD.

When too much BOD or COD is discharged into surface waters, the oxygen

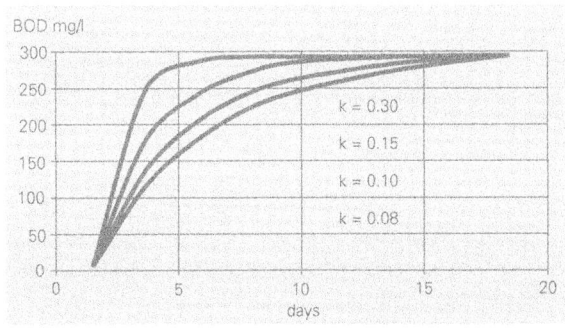

figure 8_3:
BOD-removal rates are expressed by rate constants (k) which depend on wastewater properties, temperature and treatment-plant characteristics. The curve shows the BOD-removal rates at 20°C. The value after 5 days is known as BOD_5

present in that water is consumed for the decomposition of the pollutants and, thus, is no longer available to support aquatic life. Effluent standards for discharge into receiving waters may tolerate 30 to 70mg/l BOD and 100 to 200mg/l COD.

Wastewater analysis sometimes states the total organic carbon (TOC). This indicates how much of the COD can be attributed to carbon only. In designing DEWATS, knowledge of BOD or COD is sufficient; TOC is of no practical concern.

toxic metal	concentration mg/l
Cr	28–200
Ni	50–200
Cu	5–100
Zn	3–100
Cd	70
Pb	8–30
Na	5,000–14,000
K	2,500–5,000
Ca	2,500–7,000
Mg	1,000–1,500

Table 14:
Concentration o
toxic substances
which inhibit an
robic digestion.
Source: Mudrak
Kunst, 1991

Nitrogen (N)

Most of the nitrogen in human excreta is contained in water-soluble urea. The types of nitrogen compounds found in wastewater are good indicators for which treatment steps are currently happening or have happened. Nitrogen is a major component of proteins (albumen). A high percentage of albuminoid nitrogen indicates fresh wastewater. During decomposition, when large protein molecules are broken up into smaller molecules, nitrogen is found in the form of free ammonia (NH_3, toxic for fish). However, ammonia dissolves in water and forms ammonium ions (NH_4^+) at low pH levels. At a pH level above 7, NH_4^+ transforms to NH_3. There is always a mass balance between NH_3 and NH_4. NH_3 evaporates into the atmosphere, which leads to unwanted nitrogen losses, if the treated wastewater is intended for irrigation (see section 7.1.5, page 141). Ammonium further oxidises to nitrite (NO_2^-, toxic) and finally to nitrate (NO_3^-, not toxic for fish).

From the chemical symbol, it is evident that ammonia (or ammonium) will consume oxygen to form nitrate, the most stable end-product. The albuminoid and the ammonia nitrogen together form the organic nitrogen, also called Kjeldahl-N (N_{kjel}). The total nitrogen (N_{total}) is composed of N_{kjel} (not oxidised N) and nitrate-N (oxidised N).

Pure nitrogen (N_2) is formed when oxygen is separated from NO_3 ions to oxidise organic matter. Since pure nitrogen hardly dissolves in water, it is released immediately into the atmosphere, an attribute used to remove nitrogen from wastewater in the process of denitrification. Nitrification (under aerobic conditions) followed by denitrification (under anoxic conditions) is the usual process of removing nitrogen from wastewater.

For optimum growth of micro-organisms, untreated wastewater should have a BOD/N relation of 15 to 30. Nitrogen is normally not monitored in the effluent of smaller plants. Discharge standards for the effluent of larger plants permit 10 to 20 mg/l of N_{kjel}-N.

Phosphorus (P)

Phosphorus (P) is an important parameter for planning the treatment of unknown wastewater, especially in relation to BOD, nitrogen or sulphur. Microbial growth demands approximate ratios of BOD/P and N/P of 100 and 5, respectively. Insufficient amounts of phosphorus lead to lower microbial activity and, therefore poorer removal of COD (BOD).

High phosphorus content in the effluent leads to water pollution by algae growth. However, since very little phosphorus is removed in DEWATS it is the least important parameter to the designing engineer. Discharge standards for larger plants allow P in the range of 1 to 5 mg/l.

Temperature

Temperature is an important parameter, as warmer conditions promote microbial growth. Anaerobic digestion requires minimal temperatures of 10°C; temperatures between 18 and 25°C are good, 25 to 35°C are ideal. Anaerobic processes are more sensitive to low temperatures than aerobic ones because the micro-organisms achieve lower energy gains for themselves, through the production of biogas, as an oxidiseable, energy-rich end-product. The ambient temperatures in tropical and subtropical zones are ideal for anaerobic treatment which is the basis for DEWATS.

Higher temperatures are also favourable for the growth of aerobic bacteria, but disadvantageous for oxygen transfer (Picture 7_8). A warmer environment reduces the capability of water to absorb oxygen from the air. This is the reason why ponds may become anaerobic at the height of summer.

pH-value

The pH-value indicates whether a liquid is acidic or alkaline. The scientific definition of the pH is rather complicated and of no interest to practical engineering (it indicates the H-ion concentration). Pure water has a pH of 7, which is considered to be neutral. An effluent of neutral pH indicates optimal treatment performance. Wastewater with a pH below 4 to 5 (acidic) and above 9 (alkaline) is difficult to treat; mixing tanks may be required to buffer or balance the pH level. In the case of a high pH, ammonia-N dominates, whereas ammonium-N is prevalent at low pH-values.

Volatile fatty acids

Volatile fatty acids (VFA) are used as a parameter to check the state of the digestion process. A high concentration of VFA always coincides with a low pH. Fatty acids are produced at an early stage of digestion. Too high a concentration of fatty acids indicates that the second stage of digestion, which breaks up the fatty acids, is not keeping pace with acidification. This indicates that the retention time is either too short or that the organic pollution load on the treatment system is too high. Values of VFA concentrations inside the digester in the range of BOD inflow concentration values indicate a stable anaerobic process.

Dissolved oxygen

Dissolved oxygen (DO) describes the concentration of oxygen gas that is dissolved in water. The parameter indicates the potential for aerobic treatment and is usually applied to assess the quality of surface waters. DO is vital to support aquatic life: most species of fish require a minimum of 4 to 5 mg/l DO for survival and breeding.

Treatment in DEWATS

Pathogens

The World Health Organisation (WHO) distinguishes between high-risk transmission of intestinal parasites (helminths eggs), and lower-risk transmission of diseases caused by pathogenic bacteria. Indicators for these risks are the number of helminths eggs and the number of faecal coliforms per volume of effluent, respectively. For uncontrolled irrigation less than 10,000 e-coli per litre and less than 1 helminth egg per litre is permitted by the WHO standard. E-coli bacteria are not pathogenic, but are a good indicator of faecal bacteria. Regardless of the number of ova, bacteria or viruses, wastewater is generally unsafe to humans.

Organism	Disease/symptoms
Virus (lowest frequency of infection)	
polio virus	poliomyelitis
coxsackie virus	meningitis, pneumonia, hepatitis, fever, common colds, etc.
echo virus	meningitis, paralysis, encephalitis, fever, common colds, diarrhoea, etc.
hepatitis A virus	infectious hepatitis
rota virus	acute gastroenteritis with severe diarrhoea
norwalk agents	epidemic gastroenteritis with severe diarrhoea
reo virus	respiratory infections, gastroenteritis
Bacteria (low frequency of infection)	
salmonella spp.	salmonellosis (food poisoning), typhoid fever
shigella spp.	bacillary dysentry
yersinia spp.	acute gastroenteritis, diarrhoea, abdominal pain
vibro cholerae	cholera
campylobacter jejuni	gastroenteritis
escherichia coli	gastroenteritis
Helminth worms (high frequency of infection)	
ascari lumbrocoides	digestive disturbance, abdominal pain, vomiting, restlessness
ascaris suum	coughing, chest pain, fever
trichuris trichiura	abdominal pain, diarrhea, anaemia, weight loss
toxocara canis	fever, abdominal discomfort, muscle aches, neurological symptoms
taenia saginata	nervousness, insomnia, anorexia, abdominal pain, digestive distrubance
taenia solium	nervousness, insomnia, anorexia, abdominal pain, digestive distrubance
necator americanus	hookworm disease
hymenolepsis nana	taeniasis
Protozoa (mixed frequency of infection)	
cryptosporidium	gastroenteritis
entmoeba histolytica	acute enteritis
giardia lamblia	giardiasis, diarrhoea, abdominal cramps, weight loss
balantidium coli	diarrhea, dysentery
toxoplasma gondii	toxoplasmosis

Table 15: Wastewater transmitted diseases and their symptoms

Exact pathogen counts are of limited importance for DEWATS design. Bacterial or helminth counts are important when wastewater is discharged into surface waters, which are used for bathing, washing, or irrigation.

Domestic wastewater and effluents from meat-processing plants and slaughterhouses, which carry the risk of transmitting blood-borne diseases, such as hepatitis, are particularly dangerous. The handling and discharge of such effluents may demand special precautions.

8.1.2 Dimensioning parameters

Hydraulic load

The hydraulic load is the most common parameter for calculating reactor volumes. It describes the volume of wastewater to be applied per volume of reactor, or per surface area of filter, within a given time. The usual dimension for the hydraulic load of reactors is $m^3/(m^3 \times d)$, meaning that $1m^3$ of wastewater is applied per $1m^3$ of reactor volume per day. The reciprocal value denotes the hydraulic retention time (HRT). For example, $1m^3/d$ wastewater on $3m^3$ of reactor volume results in a hydraulic load of $0.33m^3/(m^3 \times d)$, which is equal to a hydraulic retention time of three days ($3m^3$ volume/$1m^3$ of water per day).

The hydraulic retention time (HRT) gives a relation of volumes of feed in an empty reactor. It does not, for example, distinguish between sludge and liquid. The hydraulic retention time of a septic tank states nothing about the fraction of wastewater, which stays inside the tank for longer, nor does it say anything about the time that the bottom sludge has for digestion. In the case of vessels filled with filter media, the actual hydraulic retention time depends on the pore space of the media. For example, certain gravel consists of 60% stones and 40% pore space between the stones. A retention time of 24h per gross reactor volume is thereby reduced to 40%, which gives a net HRT of 9.6h.

For groundfilters and ponds, the hydraulic loading rates may be expressed in $m^3/(ha \times d)$, $m^3/(m^2 \times d)$ or $l/(m^2 \times d)$. Alternatively, the value may be stated in cm or m height of water cover on a horizontal surface. For example, 150 litres of water applied per square meter of land is equal to $0.15m^3/m^2$, which in turn is equal to 0.15m or 15cm hydraulic load.

Hydraulic loading rates are also responsible for the flow rate (velocity) inside the reactor. This is of particular interest in the case of up-flow reactors, like UASB or baffled reactors, where the up-flow velocity of liquid must be lower than the settling velocity of sludge particles. In such cases, the daily flow must be divided by the hours of actual flow (peak-hour flow rate). For calculating the velocity in an up-flow reactor, the wastewater flow per hour is divided by the surface area of the respective chamber (v = Q/A; velocity of flow equals flow divided by area). When splitting one large reactor into several chambers in series, it must be considered that the up-flow velocity in each chamber is greater than in the original large reactor. This is due to the fact that the flow rate per hour remains the same, while the area through which the flow passes is reduced to individual chambers. The necessity to keep velocity low, therefore, can lead to relatively large digester volumes, especially in anaerobic baffled reactor.

Organic load

For strong wastewater, the organic loading rate – and not the hydraulic loading rate – becomes the determining design parameter. In the case of tanks and deep anaerobic ponds, the calculation is done in grams or kilograms of BOD_5 (or COD) per m^3 digester volume per day. For shallow aerobic ponds, organic loading is related to the surface area using the dimensions grams or kilograms of BOD_5 (or COD) per m^2 or ha per day.

Table 16: Organic-loading rates and removal efficiencies of various treatment systems. Sources: mixed

typical values	aerobic pond	maturation pond	water hyacinth pond	anaerobic pond	anaerobic filter	baffled reactor
BOD_5 kg/m^3*d	0.11	0.01	0.07	0.3-1.2	4.00	6.00
BOD_5 removal	85%	70%	85%	70%	85%	85%
temperature optimum	20°C	20°C	20°C	30°C	30°C	30°C

The permitted organic loading rate is influenced by the time needed by the various kinds of micro-organisms for their specific metabolism under the given conditions (often expressed as rate constant k). This, in turn is, influenced by the kind of reactor, the reactor temperature and the kind of wastewater. Easy-to-degrade substrate can be fed at higher loading rates because the micro-organisms involved multiply fast and consume organic matter quickly. For difficult-to-degrade substrate, some of the micro-organisms require longer contact times.

Excessive loading rates can lead to "poisoning" and the process collapsing because end-products from one step of fermentation cannot be consumed by the ensuing group of micro-organisms. In anaerobic digestion, for example, overloading leads to acidification of the substrate, preventing final methanisation.

At very low loading rates, almost no sludge is produced because the micro-organisms "eat each other" for want of feed (autolysis). Consequently, incoming wastewater is not met by sufficient micro-organisms for decomposition. Although low organic loading rates do not destabilise the process, they do reduce overall treatment efficiency.

Sludge volume

The volume of sludge is an important parameter for designing sedimentation tanks and digesters. This is because the accumulating sludge occupies tank volume that must be added to the required reactor volume. The amount of biological sludge production is directly related to the amount of BOD removed which, however, depends on the decomposition process. Aerobic digestion produces more sludge than anaerobic fermentation. In addition to the biological sludge, primary sludge consists partly of settled solids, which are already mineralised.

	mineral dry matter		organic dry matter		total dry matter		BOD_5	
	g/cap.*d	g/m³	g/cap.*d	g/m³	g/cap.*d	g/m³	g/cap.*d	g/m³
settleable solids	20	100	30	150	50	250	20	100
suspended solids	5	25	10	50	15	75	10	50
dissolved solids	75	375	50	250	125	625	30	150
Total	100	500	90	450	190	950	60	300

Table 17:
Average distribution of solids in domestic wastewater in Germany.
Source: Imhoff, 19

Large, conventional sewage-treatment works remove sludge continuously and often under water – producing a very liquid sludge with a low, total solid content of between 1 and 5%. In DEWATS, the sludge remains inside the tank for at least one year, where it decomposes under anaerobic conditions and undergoes further volume reduction, as it compacts under its own weight with time.

Although the literature varies widely, it can be assumed that approximately 0.005 litres of sludge per gram BODremoved accumulate in the primary treatment step of DEWATS, including a certain percentage of mineral settleable particles. There is sludge accumulation in secondary treatment as not all digested organic matter accumulates as settleable sludge, and mineral particles have already been removed. A sludge value of 0.0075 litres per gram BODremoved can be assumed for oxidation ponds, taking additional sludge from algae into account. The above figures are estimates for "modern" domestic wastewater as described in Table 18. True sludge production is influenced by the wastewater's settling properties, ratio of organic and mineral matter content and physical boundary conditions. Further details on sludge handling and treatment can be found in section 11.3.

Table 18:
Properties of primary sludge
Source: Metcalf&Eddy, 1996

Properties of sludge from primary sedimentation		
specific grafity of solids kg/l	specific grafity of sludge kg/l	dry solids g/m^3
1.4	1.02	150.6

Additional benefits of wastewater treatment

The possible additional benefits of wastewater treatment should be considered at an early stage of planning so that it can be incorporated into the design.

Where and how the treated effluent is disposed or used affects the form of treatment that is required. While the removal of nutrients may be beneficial for the discharge into open water bodies or groundwater, it is counterproductive for reuse in agricultural irrigation. Reuse in agriculture, on the other hand, results in higher hygienic-treatment demand. More extensive treatment or dilution with fresh river-water might also be necessary to allow fish farming.

Other possible benefits from wastewater treatment, like biogas production, also restrict the choice of treatment methods, and influence investment and maintenance costs as well as amortisation.

Picture 8_4:
Sludge from a septic tank in India – the black colour the sludge indicate anaerobic conditio

Technical components

This chapter introduces the technical-treatment components of DEWATS, which correspond to the DEWATS criteria defined in chapter 7.

After a brief overview and comparison of the different technologies, detailed sections on each component explain the specifics of design, applied-treatment processes, and start-up considerations as well as operation and maintenance procedures.

9.1 Overview of DEWATS components

DEWATS is based on four treatment systems:
- sedimentation and primary treatment in sedimentation ponds, septic tanks, fully mixed digesters or Imhoff tanks
- secondary anaerobic treatment in baffled reactors (baffled septic tanks) or fixed-bed filters
- secondary and tertiary aerobic/anaerobic treatment in constructed wetlands (subsurface flow filters)
- secondary and tertiary aerobic/anaerobic treatment in ponds

Components are combined in accordance with the wastewater influent and the required effluent quality. Hybrid systems or a combination of secondary on-site treatment and tertiary co-operative treatment is also possible.

The following treatment components are discussed in further detail in the ensuing chapters:
Grease traps and grit chambers are beneficial for wastewater from canteens and certain industries. Short retention times prevent the settling of biodegradable solids. Grit and grease must be removed frequently.

Septic tanks are the most common form of treatment. The robust system provides a combination of mechanical treatment through sedimentation and biological degradation of settled organic solids. Septic tanks are used for wastewater with a high percentage of settleable solids, typically effluent from domestic sources.

Fully mixed digesters provide anaerobic treatment of wastewater with higher organic load, while serving as a settler in a combined system. In the process, biogas is produced as a useful by-product.

Imhoff tanks are slightly more complicated to construct than septic tanks, but provide a fresher effluent when de-sludged frequently. Imhoff tanks are preferred when post-treatment takes place near residential houses, in open ponds or constructed wetlands of vertical flow type.

Anaerobic baffled reactors or baffled septic tanks function as multi-chamber septic tanks. They increase biological degradation by forcing the wastewater through active sludge beneath chamber-separating baffles. All baffled reactors are suitable for all kinds of wastewater, they are most appropriate for wastewater with a high percentage of non-settleable suspended solids and narrow COD/BOD ratio.

Anaerobic filters combine mechanical solids-removal with digestion of dissolved organics. By providing filter surfaces for biological activity, increased contact between new wastewater and active micro-organisms results in effective digestion. Anaerobic filters are used for wastewater with a low percentage of suspended solids (for example, after primary treatment in septic tanks), and narrow COD/BOD ratio. Upstream Anaerobic Sludge Blanket (UASB) reactors utilise a floating sludge blanket as a biologically active filter medium.

Technical components

Trickling filters treat wastewater aerobically by letting it trickle over biologically active filter surfaces.

Horizontal gravel filters are sub-surface, flow constructed wetlands, which provide effective, facultative treatment and filtration, while allowing for appealing landscaping. Constructed wetlands are used for wastewater with a low percentage of suspended solids and COD concentrations below 500mg/l.

Pond systems are the ideal form of DEWATS treatment – if the required space is available. Anaerobic ponds are deep and highly loaded with organics. Depending on the retention time, digestion of sludge only or the complete wastewater is possible. Facultative and anaerobic ponds may be charged with strong wastewater, however, bad odour cannot be avoided reliably with high loading rates. Aerobic ponds are large and shallow – they provide oxygen via the pond surface for aerobic treatment. Wastewater for treatment in aerobic ponds should have a BOD_5 content below 300mg/l. Pond systems can be combined with certain types of vegetation, creating aquatic plant systems with additional benefits.

Special provisions are usually required for the treatment of industrial wastewater before standardised DEWATS designs can be applied. These may include open settlers for the daily removal of fruit waste from canning factories, buffer tanks for mixing varying flows from milk-processing plants, or grease traps or neutralisation pits to balance the pH of the influent. In these cases, standard DEWATS components are applicable only after such pre-treatment steps have been taken.

Despite their reliability and impressive treatment performance, such well-known and proven systems as UASB, trickling and vertical filters, rotating discs, etc. are not considered to be DEWATS because they require careful and skilled attendance.

Most treatment processes applied in conventional, large-scale treatment plants do not meet the DEWATS criteria. The activated-sludge process, the fluidised-bed reactor, aerated or chemical flocculation and all kinds of controlled re-circulation of wastewater fall within this category. Regular or continuous re-circulation might be acceptable if the pumps that are used cannot be switched off because they also act as transportation pumps.

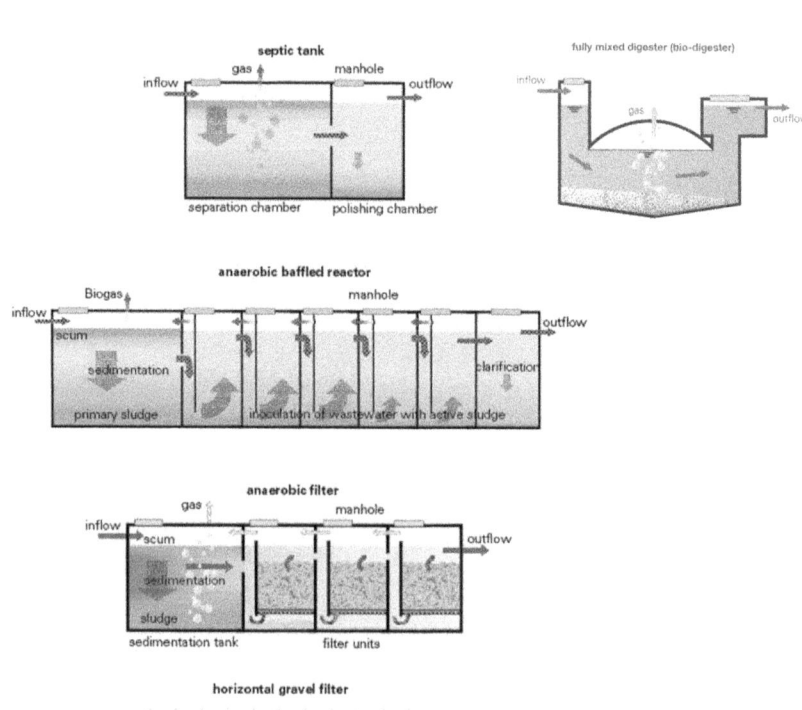

Picture 9_1:
Treatment system considered to be suitable for decentralised dissemination

Technical components

type	kind of treatment	used for type of wastewater	advantages	disadvantages
septic tank	sedimentation, sludge stabilisation	wastewater with settleable solids, especially domestic	simple, durable, little space because of being underground	low treatment efficiency, effluent not odourless
fully mixed digester	sedimentation, sludge stabilisation	concentrated organic wastewater – e.g. agro-industrial – with settleable solids	access to renewable source of energy (biogas)	less simple than septic tank; special skills needed for gas-tight dome construction
Imhoff tank	sedimentation, sludge stabilisation	wastewater with settleable solids, especially domestic	durable, little space because of being underground, odourless effluent	less simple than septic tank, needs very regular desludging
anaerobic baffled reactor	anaerobic degradation of suspended and dissolved solids	pre-settled domestic and industrial wastewater with narrow COD/BOD ratio, suitable for strong industrial wastewater	simple and durable, high treatment efficiency, little permanent space required because of being underground, hardly any blockage, relatively cheap compared to anaerobic filter	requires larger space for construction, less efficient with weak wastewater, longer start-up phase than anaerobic filter
anaerobic filter	anaerobic degradation of suspended and dissolved solids	pre-settled domestic and industrial wastewater with narrow COD/BOD ratio	simple and fairly durable if well constructed and wastewater has been properly pre-treated, high treatment efficiency, little permanent space required because of being underground	costly to construct because of special filter material, blockage of filter possible, effluent smells slightly despite high treatment efficiency
horizontal gravel filter	aerobic-facultative-anaerobic degradation of dissolved and fine suspended solids, pathogen removal	suitable for domestic and weak industrial wastewater where settleable solids and most suspended solids are already removed by pre-treatment	high treatment efficiency when properly constructed, pleasant landscaping possible, no wastewater above ground, can be cheap to construct if filter material is available at site, no nuisance of odour	high permanent-space requirement, costly if right quality of gravel not available, great knowledge and care required during construction, intensive maintenance and supervision during first 1-2 years
anaerobic pond	sedimentation, anaerobic degradation and sludge stabilisation	strong and medium industrial wastewater	simple in construction, flexible in respect to degree of treatment, little maintenace	wastewater pond occupies open land, there is always some odour, can even be stinky, mosquitoes are difficult to control
aerobic pond	aerobic degradation, pathogen removal	weak, mostly pre-treated wastewater from domestic and industrial sources	simple in construction, reliable in performance if properly dimensioned, high pathogen removal rate, can be used to create an almost natural enviroment, fish farming possible when large in size and low loaded	large permanent space requirement, mosquitoes and odour can become a nuisance if undersized near residential areas, algae can raise effluent BOD

Table 19:
Pros and Cons of DEWATS

Admittedly, these self-imposed restraints on DEWATS can, in practice, impact the quality of the effluent. But this need not be the case if there is sufficient space for the plant. Measures to discharge effluent of acceptable quality include:
- provision of sufficient space at the source of pollution
- pre-treatment at source and post treatment where sufficient land is available
- pre-treatment at source and post treatment in co-operation with others
- accepting an effluent with higher pollution load
- restricting wastewater-producing activities at this particular site
- connection to a central treatment plant via a sewage line

The permanent dilution of wastewater or the installation of a highly mechanised, "modern" treatment plant remain theoretical options – experience shows that such processes are chronically afflicted by irregular operation.

Picture 9_2:
One of too many non-aerating aerators

Space requirements

Depending on the total volume and the nature of the wastewater and its temperature, the following values may indicate permanent area requirements for setting up a treatment plant:

Septic tank, Imhoff tank:	$0.5 m^2/m^3$ daily flow
Anaerobic baffled reactor, anaerobic filter:	$1 m^2/m^3$ daily flow
Horizontal gravel filter:	$30 m^2/m^3$ daily flow
Anaerobic ponds:	$4 m^2/m^3$ daily flow
Facultative aerobic ponds:	$25 m^2/m^3$ daily flow

These values are approximations for wastewater of typical strength; land requirements increase with wastewater of higher pollution load. Land use can be minimised if closed anaerobic systems are applied, as they are usually constructed underground. The area for sludge-drying beds may require an additional 0.1 to 10m²/m³ daily flow, depending on the wastewater quality and desludging intervals.

Performance

Treatment quality depends on the nature of the influent and boundary conditions like temperature. BOD-removal rates are generally within these ranges:
25 to 50% for septic tanks and Imhoff tanks
70 to 90% for anaerobic baffled reactors and anaerobic filters
70 to 95% for horizontal gravel filter and pond systems

The treatment efficiency of the different components and the required effluent quality decide the choice of treatment system. For example, septic tanks alone are not adequate for direct discharge into surface waters, but may suit treatment on land where the groundwater table is low and odour is not likely to be a nuisance. Assuming a discharge limit of 50mg/l BOD, the anaerobic filter in combination with a septic tank may treat wastewater of 300mg/l BOD without further treatment. Stronger wastewater would require a horizontal gravel filter or pond system for final treatment. Perhaps even long-way open discharge channels are sufficient to provide the necessary additional treatment.

Based on local conditions, many other possibilities for cheaper treatment systems may exist – all options must be considered. Expert knowledge is needed to evaluate such possibilities; wastewater-sample analysis should be compulsory.

Substantial removal of nitrogen requires a mix of aerobic and anaerobic treatment, only provided by constructed wetlands and ponds. In closed anaerobic-tank systems of the DEWATS-type, nitrogen forms to ammonia. The effluent is a good fertiliser but causes algae growth and is toxic to fish if released into surface waters.

Phosphorus is a good fertiliser and, therefore, dangerous in rivers and lakes. Phosphorus removal in DEWATS is limited – as in most treatment plants. Constructed wetlands with filter media containing iron or aluminium compounds present one form of removal. Furthermore phosphorus can be accumulated by sedimentation or fixed in microbial mass, although it can hardly be removed from the sludge or be transformed into a less-harmless state.

Pathogen control

Like all other modern wastewater-treatment plants, DEWATS systems are not focused on pathogen control. Pathogen removal increases with longer retention times, but treatment plants proudly function on short HRTs.

The WHO guidelines and other independent surveys describe the transmission of worm infections as the greatest risk associated with wastewater. Worm eggs or helminths are, for the most part, removed from effluent by sedimentation and accumulate in the bottom sludge. The long retention times in septic tanks and anaerobic filters of 1 to 3 years provide sufficient protection against helminths infection; frequent sludge removal is discouraged due to increased health risks.

Although many bacteria and viruses are destroyed during treatment, the concentrations in the effluent of anaerobic filters and septic tanks are still infectious. Higher pathogen-removal rates are reported from constructed wetlands and shallow aerobic ponds; the effect is attributed to longer retention times, exposure to UV rays in ponds, and various bio-chemical interactions in constructed wetlands. The pathogen-removal rates of these systems are, in fact, higher than in conventional municipal treatment plants.

Chlorination can be used for pathogen control. Simple devices with automatic dosing may be added before final discharge. However, the use of chlorine should be limited to cases of high risk, such as hospital wastewaters during an epidemic. Permanent chlorination should be avoided because it not only kills pathogens but also destroys other bacteria and protozoa, which are responsible for the self-purification effect of receiving waters.

9.2 DEWATS Modules

9.2.1 Grease trap and grit chamber

If a septic tank is provided, DEWATS normally do not require grease traps or grit chambers for domestic wastewater. Whenever possible they should be avoided altogether because grease and grit must be removed, at least once a week. However, for canteens or certain industrial wastewaters it may be advisable to separate grit and grease before the septic tank.

The function of grease and grit chambers is comparable to that of septic tanks; light matter should float and heavy matter should sink to the bottom. The difference is that bio-degradable solids should have no time to settle. Retention times for grit chambers are short, therefore, – only about three minutes. The use of masonry structures is not appropriate, especially in the case of minor flows.

A conical trough allows slow flow at a large surface for grease floatation and fast flow at the narrow bottom, which allows only heavy and coarse grit to settle. The water surface is protected from the turbulence of the inflow by a baffle; the outlet is near the bottom.

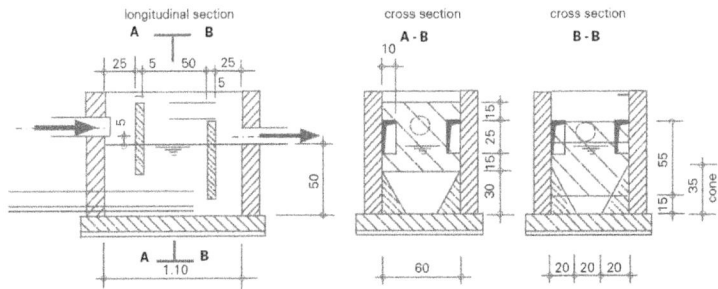

Figure 9_3:
Design principle of combined grease trap and grit chamber. Accumulating grease, oil and grit should be removed daily, or at least weekly. If this can not be assured, an oversized septic tank is preferable to receive grit and grease

9.2.2 Septic tank

The septic tank is the most common, small scale and decentralised treatment plant, worldwide. It is compact, robust and extremely efficient when compared with the cost of constructing it. It is basically a sedimentation tank in which settled sludge is stabilised by anaerobic digestion. Dissolved and suspended matter leaves the tank more or less untreated.

Two treatment principles, namely the mechanical treatment by sedimentation and the biological treatment by contact between fresh wastewater and active sludge, compete with each other in the septic tank. Optimal sedimentation takes place when the flow is smooth and undisturbed. Biological treatment is optimised by quick and intensive contact between new inflow and old sludge, particularly when the flow is turbulent. How the influent enters and flows through the tank decides which treatment effect predominates.

With smooth and undisturbed flow, the supernatant (the water remaining after settleable solids have separated) leaves the septic tank rather fresh and odourless, implying that degradation has not yet started. With turbulent flow, the degradation of suspended and dissolved solids starts immediately because of the intensive contact between fresh and already active substrate. However, as turbulence hinders sedimentation, more suspended solids are discharged with the effluent, resulting in odours because active solids, which are not completely fermented, leave the tank.

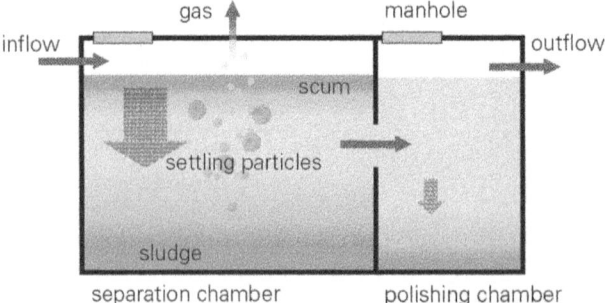

Picture 9_4:
Flow principle of the septic tank. Most sludge and scum is retained the first chamber the second chamber contains only little sludge, which allows the water flow without disturbance from rising gas bubbles

Domestic wastewater normally forms a heavy scum near the inlet. This consists of matter lighter than water, such as fat, grease, wood-chips, hair or any floating plastics. A larger portion of the floating scum also consists of sludge particles, which are released from the bottom and driven to the top by treatment gases. New sludge from below lifts the older scum particles above the water surface where they dry and become lighter. The accumulated scum must be removed regularly, at least every third year. Scum does not harm the treatment process as such, but it does occupy tank volume.

A septic tank consists of a minimum of two, sometimes three compartments. The compartment walls extend 15cm above the liquid level. They may also be used as bearing walls for the covering slab if some openings for internal gas exchange are provided.

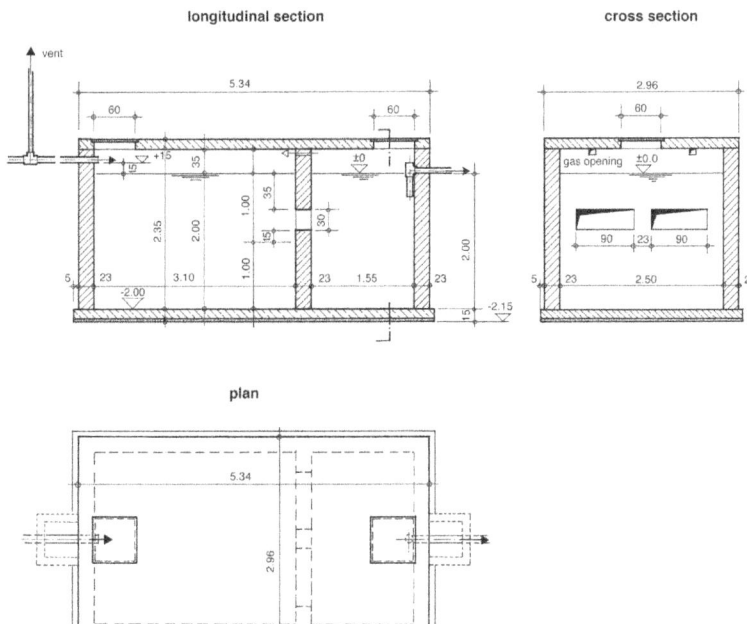

ture 9_5:
The septic tank. The dimensions have been calculated for 13m³ of domestic wastewater per day

The first compartment occupies about two-thirds of the septic-tank volume, allowing for most of the sludge and scum accumulation. The following chamber(s) are provided to calm the turbulent liquid. They are all the same size and make up the remainder of the volume. All chambers are normally the same depth. The depth from the outlet level to the bottom should be between 1.50m and 2.50m. The first chamber is sometimes deeper.

The size of the first chamber is calculated to be at least twice the accumulating sludge volume. The sludge volume depends on the settleable solids content of the influent and on desludging intervals (see picture 10_5, page 238). Most countries provide a National Standard for tank volume per domestic user.

The SS removal rate drops drastically when accumulated sludge fills more than two-thirds of the tank. This must be avoided, especially in cases where the effluent is treated further in a sand or gravel filter.

"Irregular emptying of septic tanks leads to irreversible clogging of the infiltration bed; rather than renewing the bed, most owners bypass it and divert the tank's effluent to surface drains."[33]

33 See: Alearts et a 1990

For domestic sewage, the accumulating sludge volume can be calculated with 0.1l/cap×d. When desludging intervals are longer than two years, the sludge volume may be reduced to 0.08l/cap×d, as sludge compacts with time (see Picture 10_5).

The inlet may dive down inside the tank, below the assumed lowest level of the scum – or may be above the water level when the inlet pipe is used to evacuate gas. A septic tank is basically a biogas plant, without biogas use. Gas accumulates inside the tank above the liquid, from where it should be able to escape into the air. The ventilation pipe for digester gases should end outside buildings, at an elevation above roof level. Open fire should be avoided when opening the septic tank for cleaning.

Technical components

The compartments are connected by simple wall openings situated above the highest sludge level and below the lowest level of the scum. For domestic wastewater, the top of the opening should be 30cm below outlet level, its base at least half the water depth above the floor. The openings should be equally distributed across the width of the tank, in order to minimise turbulence. A slot, spanning the full width of the tank, is ideal for reducing velocity and turbulence.

The outlet has a T-joint, the lower arm of which dives 30cm below the water level. With this design, foul gas trapped in the tank enters the sewage line from where it must be ventilated safely. If ventilation cannot be guaranteed, an elbow must to be used at the outlet to prevent the gas from entering the outlet pipe. There should be manholes in the cover slab; one each above inlet and outlet and one at each baffle wall, preferably at the inlet of each compartment. The manholes should permit water sampling from each compartment.

Septic tanks were originally designed for domestic wastewater. They are also suitable for other wastewater of similar properties, particularly those that contain a substantial portion of settleable solids.

The treatment efficiency of a septic tank ranges from 25% to 50% COD removal. It serves as rough, primary treatment, prior to secondary or even tertiary treatment. Post-treatment may be provided in ponds or ground filters. In the latter case, regular desludging of septic tanks is mandatory. A septic tank may also be integrated into an anaerobic filter or as the first section of a baffled reactor. Septic tanks are suitable as individual on-site pre-treatment units for community sewer systems because the diameter of sewerage can be smaller when settleable solids have been removed on-site.

Starting phase and maintenance

A septic tank may be used immediately; it does not require special arrangements before usage. However, sludge digestion begins only after several days. Regular desludging is required every one to three years. When removing the sludge, some immature (still-active) sludge should be left inside the tank to enable continuous decomposition of newly settling solids; it is not necessary to remove the liquid. This means, if the sludge is removed by pumping, the pump head should be brought down to the very bottom. Adequate handling and treatment of septic sludge is discussed in detail in section 11.3. The septic tank's surroundings should be kept free of plants to prevent roots from growing in the pipelines and control chambers.

Calculating dimensions

Approximately 80 to 100l should be provided per domestic user. For exact calculation or for wastewater from non-domestic sources, the formula applied in the computer spreadsheet (Table 25, page 240) may be used.

9.2.3 Fully mixed digester

The fully mixed anaerobic digester (also called bio-digester) corresponds to the biogas plants, which are often used by farming families in developing countries. It is suitable for rather "thick" and homogenous substrate like sludge from aerobic-treatment tanks or liquid animal excreta. For economic reasons, it is not suitable for weak-liquid wastewater because the total volume of wastewater must be agitated and kept inside the digester for the full retention time of 15 to 30 days. This results in larger digester volumes and higher construction costs. However, combining different waste sources or blackwater from several toilets can be considered.

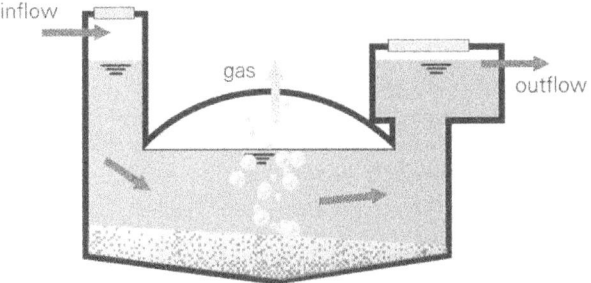

fully mixed digester (bio-digester)

Figure 9_6:
Flow principle of the fully mixed anaerobic digester. There is little sedimentation due to viscous substrate. All substrate fractions stay inside the digester for the same period of time. The position of inlet and outlet is less important with homogeneous liquid of high TS content. Small baffles may be provided to avoid short circuiting

"Thick" viscous substrates of more than 6% total solid content do not need stirring. A digester with such a substrate can be operated for many years without desludging because only grit, but hardly any sludge, settles. Moreover, all the incoming substrate leaves the reactor after digestion. Scum formation is still possible with certain substrates. Therefore, if inlet and outlet pipes are used they should be placed at middle height. In fixed-dome digesters, the outlet should be made of a vertical shaft with the opening starting immediately below the zero-line; this will allow some of the scum to discharge.

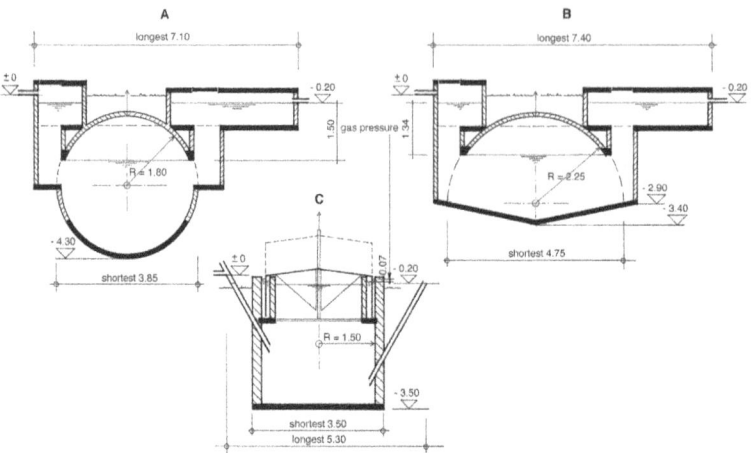

Picture 9_7:
Traditional biogas plants as fully mixed anaerobic digester. A: The ball-shaped fixed dome plant with integrated gas storage expansion chamber. B: The half-ball-shaped fixed-dome plant. C: The floating drum plant water seal. All the plants are designed for 600 litres substrate per day of organic dry-matter content, at 25°C and HRT of 25 days. The expected gas production is 8.4 m³/d. A comparison of the space requirements and gas pressure of all the plants indicates floating drum plants are preferable for high gas-production rates.

Since the fully mixed digester is only used for strong substrate, biogas production is high and can be used afterwards. In this case, the gascollector tank and the gas-storage tank must be gas-tight. The immediate gas outlet should be 30cm above substrate level. Smaller units usually use the fixed dome (hydraulic-pressure) system made out of masonry structure, while larger units store the biogas in steel-drums or plastic bags.

The choice of gas-storage system will depend on the pattern of gas utilisation. Ideally, gas production should coincide with gas consumption, in time and volume. For more details, please refer to chapter 6, Biogas utilisation. An abundance of special biogas literature is also available.

Starting phase and maintenance

Starting with some active sludge from a septic tank speeds up digestion and prevents the digester from turning sour. In the rare case of this happening, the loading rate should be reduced until the pH turns neutral. It may be necessary to remove sand and grit after several years.

Technical components

Calculating dimensions

The main parameter is the hydraulic retention time, which should not be less than 15 days in a hot climate and not less than 25 days in a moderately warm climate; a HRT of more than 60 days is required for highly pathogenic substrate. The gas-storage volume depends on daily gas use in relation to daily gas production. The storage capacity of gas for household use should exceed 65% of the daily gas production. Gas production is directly related to the organic fraction of the substrate. In practice, it is calculated as a fraction of the daily substrate that is fed. Experience indicates, for example, that 1kg fresh cattle dung diluted with 1 litre of water produces 40l of biogas. More exact calculations will be obtained by using the formulas applied in the spreadsheet Table 26, page 246.

9.2.4 Imhoff tank

Imhoff or Emscher tanks are typically used for domestic or mixed wastewater flows above 3m³/d when the effluent receives further treatment above the ground and, therefore, should not stink – as may be the case with septic tanks. The Imhoff tank effectively separates fresh influent from bottom sludge.

The tank consists of a settling compartment above the digestion chamber. Funnel-like baffle walls prevent up-flowing foul-sludge particles from mixing with the effluent and causing turbulence. The effluent remains fresh and odourless because the suspended and dissolved solids do not come into contact with the active sludge and turn sour and foul. Retention time should not be much more than 2 hours during peak flow, otherwise this effect is jeopardised.

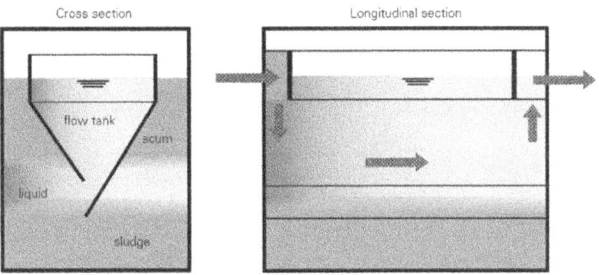

ture 9_8:
Flow principle of the Imhoff tank. The water passes quickly through the flow tank, undisturbed by rising gas bubbles; the water does not mix with "ripe" water or sludge

When sludge ferments at the bottom, the sludge particles get attached to foulgas bubbles and start floating upwards. The up-flowing sludge particles assemble outside the conical walls and form an accumulating scum layer, which grows continuously downwards. When the slots – through which settling particles should fall into the lower compartment – are closed, the treatment effect is reduced to that of a undersized septic tank. Sludge and scum, therefore, must be removed at appropriate intervals.

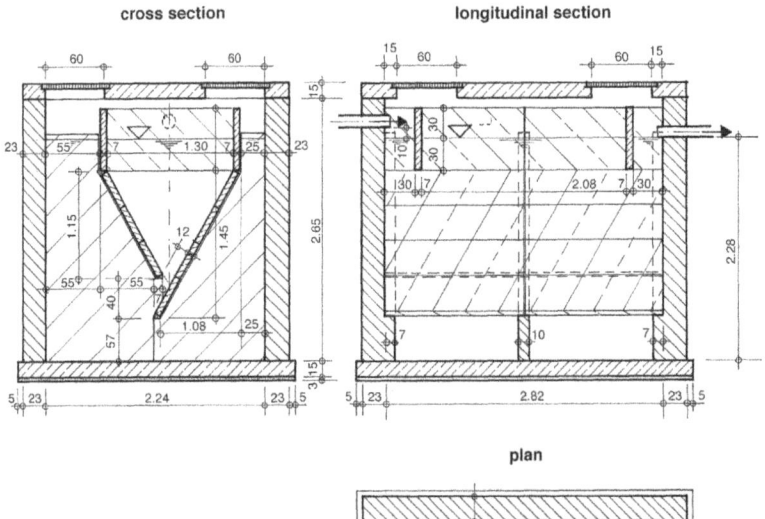

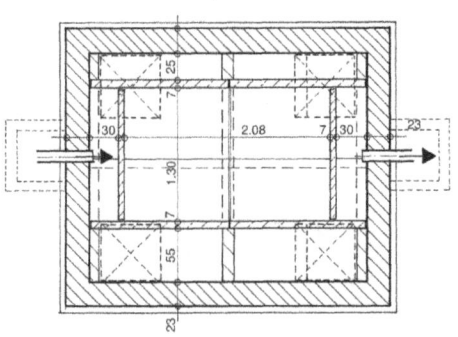

Picture 9_9:
Imhoff tank.
Dimensions have been calculated 25m³ of domestic wastewater per

The inlet and outlet pipes are the same shape as those in septic tanks. Pipe ventilation must be provided, as Imhoff tanks also produce biogas. Additional baffles to reduce velocity at the inlet – and to retain suspended matter at the outlet – are advantageous. The upper part of the funnel-shaped baffles is vertical for 30cm above and 30cm below the water surface. The shape of an Imhoff tank may be cylindrical; the funnel, however, should always be rectangular, in order to leave adequate space outside the funnel for scum removal. The funnel structure may consist of pre-fabricated ferro-cement. Treatment efficiency lies in the range of 25 to 50% COD reduction.

Starting phase and maintenance

As with septic tanks, no special start-up phase is required. Desludging is necessary at regular intervals. Sludge should be removed from the bottom of the tank by pumping or hydraulic pressure pipes, withdrawing only fully digested substrate and leaving some active sludge behind for maintaining microbial activity. Best practice in the removal, handling and treatment of sludge is discussed in detail in section 11.3.

Scum must be removed before it grows enough to close the slots between the upper and lower compartments. Should this happen, gas bubbles appearing in rows on the water surface above the slots indicate excessive scum accumulation. Scum should be removed before sludge removal; the liquid may remain inside the tank.

Calculating dimensions

The upper compartment, inside the funnel walls, should be designed for 2h HRT at peak flow, and the hydraulic load should be less than 1.5m³/h per 1m² surface area. The sludge compartment below the slots should be calculated to retain 2.5 litres of sludge per kg BOD reduced per day for short desludging intervals. For longer intervals please refer to the corresponding spreadsheet (Table 27, page 248).

For domestic wastewater and desludging intervals of one year, the upper compartment should have a volume of approximately 50l per user and the sludge compartment below the slots should have a volume of approximately 120 litres per user. This is only a rule of thumb; for more detailed calculations, or for wastewater from non-domestic sources, please refer to the spreadsheet.

9.2.5 Anaerobic baffled reactor

The anaerobic baffled reactor (ABR), also known as the "baffled septic tank", can be considered as the DEWATS version of the UASB system. It is, in fact, a combination of several anaerobic-process principles: the septic tank, the fluidised bed reactor and the UASB.

The up-flow velocity of the baffled reactor, which should never be more than 1m/h, limits its design. Based on a given hydraulic retention time, the up-flow velocity increases in direct relation to the reactor height. The reactor height, therefore, can not be used as a variable parameter to achieve the required HRT. The limited upstream velocity results in large but shallow tanks, making the system uneconomical for larger plants. This is why baffled reactors are not very well-known or properly researched.

However, the anaerobic baffled reactor is ideal for DEWATS because it is simple to build and simple to operate. Hydraulic and organic shock loads have little effect on treatment efficiency.

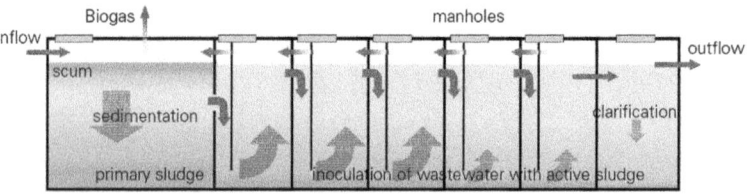

Picture 9_10:
Flow principle of an ABR. Incoming waste-water is forced to pass through a sludge blanket of active micro-organisms in each compartment. The settler the front prevents larger solids from entering the baffle section

The main difference to the UASB is that it is not necessary for the sludge blanket to float; it may rest at the bottom. Three-phase separators are also unnecessary because active sludge washed out from one chamber is trapped in the next. Tanks in series also help to digest difficult degradable substances, predominantly in the later chambers after easily degradable matter has been digested in the earlier ones. The anaerobic baffled reactor consists of at least four chambers in series. But practical experience shows that treatment efficiency does not increase with more than six chambers. The last chamber can incorporate a filter in its upper part, in order to retain remaining solid particles; alternatively, a settler for post-treatment can follow the baffled reactor (Picture 9_36).

Technical components

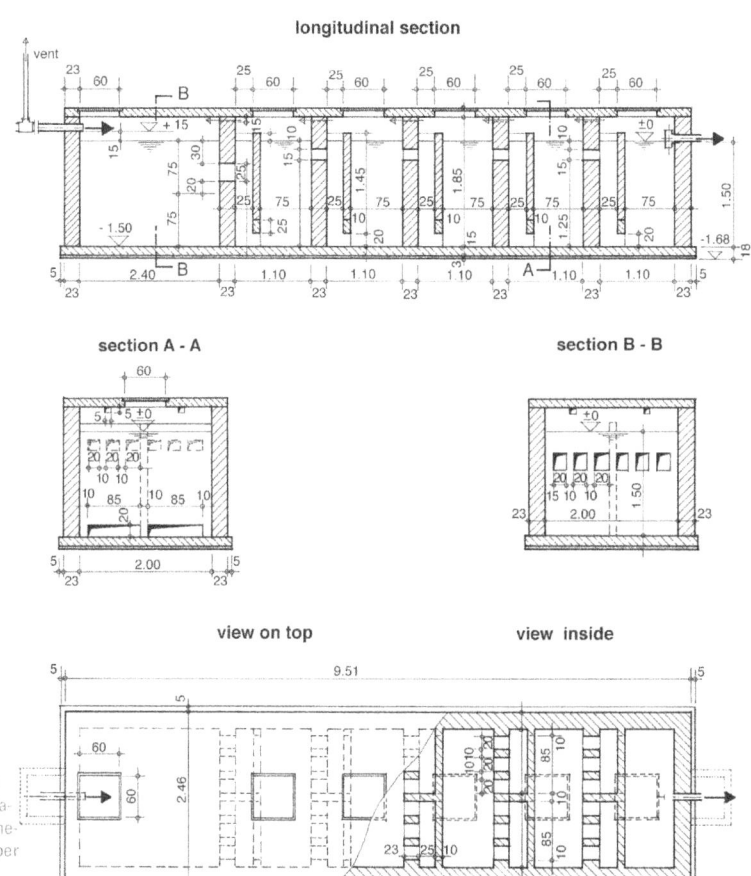

Figure 9_11:
ABR Dimensions have been calculated for 25m³ domestic wastewater per day

Equal distribution of inflow, and extensive contact between new and old substrate are important process features. Unlike in the Imhoff tank, the fresh influent is immediately mixed – and, thereby, inoculated – with the active sludge in the reactor, to begin digestion. The wastewater flows from bottom to top with the effect that sludge particles settle against the up-stream of the liquid, providing intensive contact between resident sludge and newly incoming liquid.

The DEWATS version of the anaerobic baffled reactor does not have a rack or screen. A settling chamber is used to separate the larger solids before the wastewater continues to a series of up-flow chambers. Between chambers the water flow is directed to the bottom of the next chamber by baffle walls that form a down-shaft, or by down-pipes that are placed on the partition walls. Although down-pipes reduce the total digester length (and the cost) down-shafts are preferable because of better flow distribution.

The wastewater that enters a tank should be distributed over the floor area as evenly as possible. This is facilitated by relatively short compartments (length < 50% to 60% of the height) or, in the case of down-pipes, a distance of less than 75cm between pipes. In larger plants, when longer compartments are required, down-pipe outlets (as well as down-shafts) should reach to the centre of the floor area.

The outlet of each chamber (particularly the last one) should be placed slightly below the water surface to retain possible scum. Although not common practice, baffled reactors can be equipped with three-phase separators – in the form of slanting baffles in the upper third of the tank.

The anaerobic baffled reactor is suitable for treating all kinds of wastewater with BOD above BOD < 150mg/l. Although its efficiency increases with higher organic loading, it is also well-suited for domestic wastewater. There is relatively little experience with baffled reactors because the system is only used in smaller units. As a highly efficient modification of the less-efficient septic tank, baffled reactors combine simple and efficient operation with easy, low-cost construction. Treatment performance is in the range of 65% to 90% COD (70% to 95% BOD) removal. However, three months are required for maturation.

Technical components

Starting phase and maintenance

Treatment performance depends on the availability of active microbial mass. Inoculation with old sludge from septic tanks shortens the start-up phase. In principle, it is advantageous to start with only a quarter of the daily flow and with a slightly stronger wastewater. The loading rate should increase slowly over three months. This provides micro-organisms with enough time to multiply before suspended solids are washed out. Starting with the full hydraulic load from the beginning severely delays maturation.

Like regular septic tanks, sludge must be removed at regular intervals, leaving some sludge to ensure continuous treatment efficiency. More sludge accumulates in the front than in the rear compartments. Adequate removal, handling and treatment of sludge is discussed in detail in section 11.3.

Calculating dimensions

The up-flow should not exceed 1.0m/h. This is the most crucial parameter for dimensioning, especially with high hydraulic loading. The organic load should be below 3.0kg $COD/m^3 \times d$. Higher loading rates are only possible at higher temperatures and for easily degradeable substrate. The HRT of the liquid fraction (i.e. above the sludge volume) should not be less than eight hours. Sludge-storage volume should be provided for $4l/m^3$ BODinflow to the settler and $1.4l/m^3$ BODremoved in the upstream tanks. For exact calculation use the formula applied in the spreadsheet (Table 28, page 252).

Picture 9_12:
Baffled reactor under construction at a factory in India

9.2.6 Anaerobic filter

The dominant principle of both the septic and Imhoff tanks is sedimentation combined with sludge digestion. The anaerobic filter, also known as a fixed-bed or fixed-film reactor, is different in that it also includes the treatment of non-settleable and dissolved solids by bringing them into close contact with a surplus of active microbial mass.

"Hungry" micro-organisms digest the dispersed or dissolved organic matter within a short retention time. Most of the micro-organisms are immobile; they attach themselves to solid particles or, for example, the reactor walls. Filter material, such as gravel, rocks, cinder or specially formed plastic shapes, provide additional surface area for them to settle. By forcing the fresh wastewater to flow through this material, intensive contact with active micro-organisms is established; the larger the surface for microbial growth, the quicker the digestion. Good filter material provides 90 to 300m² surface area per m³ of occupied reactor volume. Rough surfaces provide a larger area, at least in the starting phase; the microbial "lawn" or "film" that grows on the filter mass quickly closes the smaller grooves and holes.

Picture 9_13:
Floating filter bal made of plastic. When the film of micro-organisms becomes too hea the balls turn ove and discharge the load. The filter me dium has success fully been used fc tofu wastewater HRIEE in Zheijian Province in China

Technical components

Anaerobic filters are very reliable and robust. Experience shows, however, that between 25 to 30% of the total filter mass may be inactive due to clogging. While a cinder or rock filter may not completely become, reduced treatment blocked efficiency is indicative of clogging in some parts. Sand or gravel filters may block up completely due to smaller pore size.

Clogging happens when wastewater finds a channelled way through just a few open pores; eventually, the lessused voids clog and higher flow velocities occur in the few remaining open. This leads to reduced retention time and active micro-organisms are washed away.

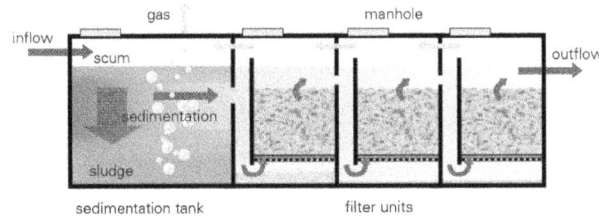

ture 9_14:
Flow principle of anaerobic up-flow filter. Most suspended solids should be retained in the septic tank. Filter micro-organisms consume dissolved and dispersed solids. Anaerobic filters may also be designed for down-flow

When the microbial film becomes too thick it must be removed. This may be done either by back-washing or by removing the filter mass for cleaning outside the reactor.

The treatment efficiency of well-operated anaerobic filters ranges between 70 to 90% BOD removal. They are suitable for domestic wastewater and all industrial wastewater with low suspended-solids content. Pre-treatment in settlers or septic tanks may be necessary to eliminate larger solids before the wastewater enters the filter.

Anaerobic filters may be operated as down-flow or up-flow systems. The up-flow system is normally preferred because there is less risk of washing out active micro-organisms. On the other hand, flushing the filter – or cleaning – is easier in down-flow systems. A combination of up-flow and down-flow chambers is also possible.

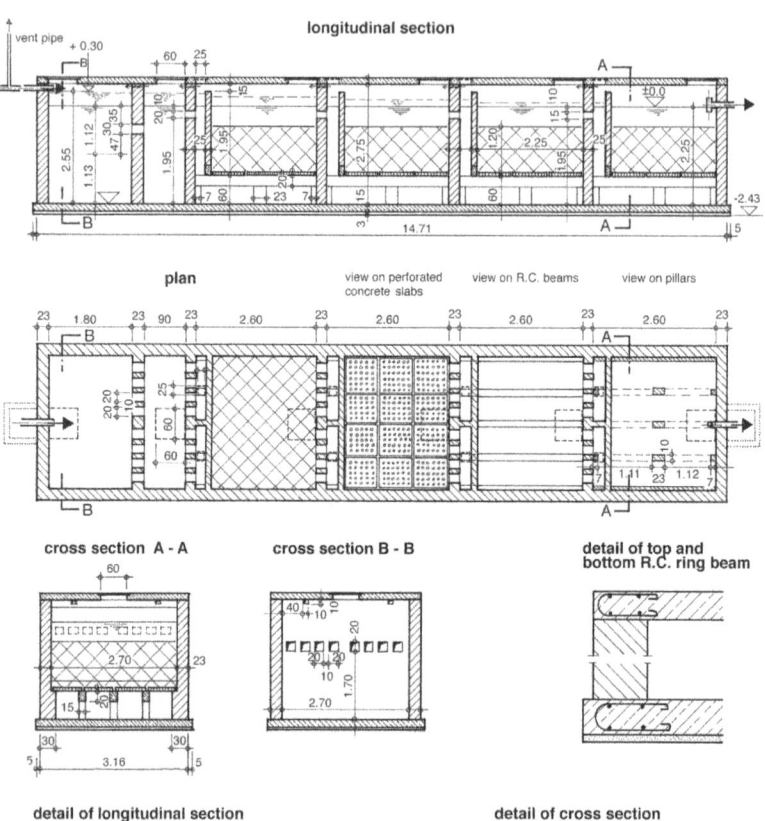

Picture 9_15: Anaerobic filter – dimensions have been calculated for 25m³ domesti[c] wastewater per d

Technical components

An important design criterion is the equal distribution of wastewater across the filter area. Equal distribution is facilitated by providing adequate free-flow space across the full width before and after the filter. This is why full-width down-flow shafts are preferred to down-flow pipes. The length of the filter chamber should not be greater than the depth of the water.

For smaller and simple structures, the filter mass consists of cinder (5 to 15cm in diameter) or rocks (5 to 10cm in diameter), which are – bedded on perforated-concrete slabs. The filter starts with a layer of large rocks at the bottom. The slabs rests on beams, which are parallel to the direction of flow, approximately 50 to 60cm above the ground slab. Pipes of at least 15cm diameter, or down-shafts over the full width, permit desludging at the bottom with the help of pumps from the top. In case the sludge-drying beds are located directly beside the filter, sludge may also be drawn via hydraulic-pressure pipes. Head losses of 30 to 50cm have to be considered.

Biogas utilisation may be considered in case of BOD concentration > 1,000mg/l; this requires completely gas-tight construction and provisions for collection, storage and use.

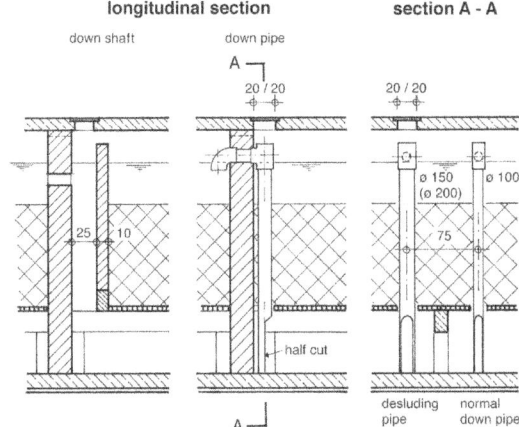

Figure 9_16: down-shaft and down-pipes – both systems may be applied alternatively in anaerobic filters and ABRs

Starting phase and maintenance

Since the treatment process depends on a surplus of active microbial mass, active sludge (for example, from septic tanks) should be sprayed on the filter material before continuous operation is started. If possible, start with only a quarter of the daily flow, and increase the flow slowly over three months. As this might not be possible in practice, treatment is unlikely to be operating at full capacity until approximately six to nine months later.

As with septic tanks, desludging should be done at regular intervals. Where possible, the filter should be back-washed before sludge removal. Adequate removal, handling and treatment of sludge is discussed in detail in section 11.3. The filter should be cleaned when efficiency declines.

Calculating dimensions

Organic-load limits between range 4 to 5kg $COD/m^3 \times d$. The hydraulic retention time compared to the tank volume should range between one and a half and two days. For exact calculation, please refer to the spreadsheet (Table 29, page 257). For domestic wastewater, constructed gross digester volume (voids plus filter mass) may be estimated at $0.5m^3$/capita; for smaller units it is closer to $1m^3$/capita.

9.2.7. Planted soil filters

Three basic-treatment systems are referred to as planted soil filters:
- overland-treatment systems
- vertical-flow filters and
- horizontal-flow filters

In overland treatment the water is distributed on carefully contoured land by sprinklers. As the system requires permanent attendance and maintenance is not considered a component of DEWATS.

In vertical-filter treatment (see picture below) the wastewater is alternately distributed on two or three filter beds with the help of a dosing device (similar to the trickling filter). The treatment functions, predominantly, aerobically. Although vertical filters require only about half the area of their horizontal counterparts and often achieve higher treatment efficiency, the constant operational control, need for a dosing device and strict adherence to charging intervals make vertical filters less suitable for DEWATS.

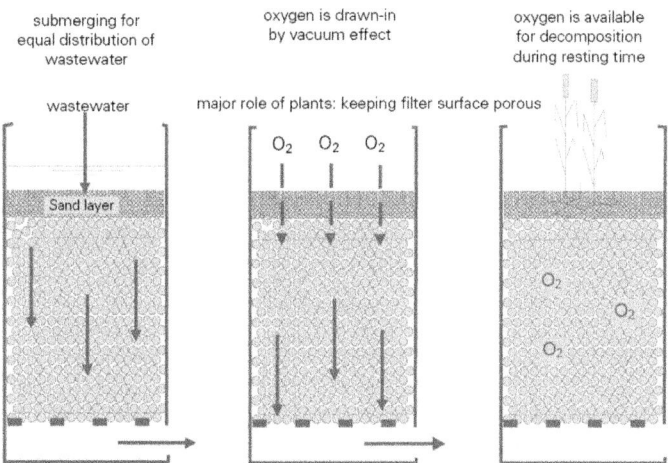

Picture 9_17:
The principle of the vertical filter

Horizontal filters comply with DEWATS criteria, as they are simple in principle and require almost no maintenance – if well-designed and constructed. Planted horizontal gravel filters – also referred to as subsurface flow wetlands (SSF) or root zone treatment plants – provide natural treatment for pre-settled wastewater of a maximum COD content of 500mg/l. They are ideal, therefore, as tertiary treatment for wastewater, which has already undergone secondary treatment in units, such as baffled reactors, anaerobic filters or biogas digesters. They are also appropriate for treating pre-settled greywater directly.

Although they don't look complicated – and are quite simple to operate, designing sand and gravel filters requires a solid understanding of the treatment process and good knowledge of the filter medium that is to be used. Before deciding on filter treatment, therefore, one should always consider the alternative: constructing wastewater ponds. Filter treatment, however, has the great advantage of keeping the wastewater below ground, thereby avoiding smells and insect breeding.

9.2.7.1 Horizontal gravel filter

Since clogging is the biggest problem with horizontal gravel filters, the wastewater must be pre-treated so that suspended solids are removed before it enters the treatment unit. When testing wastewater, after 60 minutes in an Imhoff cone the sediment should not be more than 1ml/l, and not more than 100mg SS/l for non-settling industrial wastewater. If the COD-value of settleable solids is less than 40% of the total SS-value, then many of the solids are likely to be fat in colloidal form, which can reduce the hydraulic conductivity of the filter considerably (as may be the case with dairy wastewater).

The treatment process in horizontal ground filters is complex and not yet fully understood. Unlike the vertical filter, the horizontal filter (Picture 9_18) is permanently soaked with water and operates partly aerobic (free oxygen present), partly anoxic (no free oxygen but nitrate – NO_3 – present) and partly anaerobic (no free oxygen and no nitrate present). Combined with physical-filtration processes and the influence of plantation on the biological-treatment process and oxygen intake, the interaction of the separate treatment processes is difficult to predict. There are sophisticated methods for calculating the proper dimensions and treatment characteristics of different filter media, especially in relation to their hydraulic properties. However, such calculations make sense only if the exact required parameters are known, which is hardly ever the case. Rules of thumb, intelligently chosen, are more than sufficient for smaller-sized DEWATS plants. Going beyond these experience-based figures is not advisable without previous tests.

The rules of safe design are:
- large and shallow filter-bed
- wide inlet zone
- reliable distribution of inflow over the full width of the inlet zone
- round, coarse gravel that is nearly the same size as the filter medium

Technical components

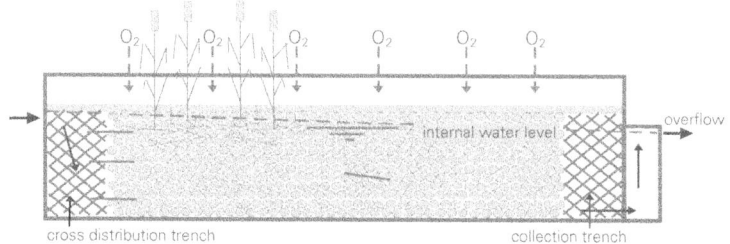

continuous oxygen supply to the upper layers only major role of plants: provide favourable environment for bacteria diversity

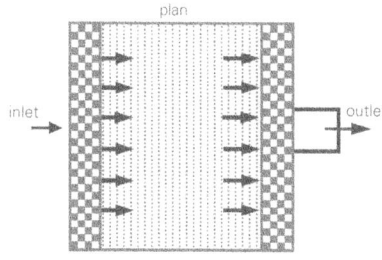

anaerobic and anoxic conditions in the lower layers

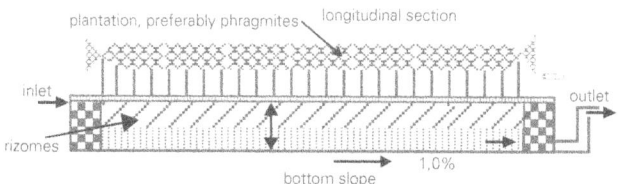

water flow in the horizontal filter

Figure 9_18:
The principle of the horizontal filter

Clogging is caused by suspended solids and by biological or mineralised sludge newly formed from the decomposition of organic matter. While large grain sizes with a high percentage of voids prevent clogging, they also reduce treatment performance. In order to utilise the full filter, the front part of the bed must have voids that are small enough to retain some of the SS, while being large enough to allow further SS removal in later parts of the bed. Round, uniform gravel of 6-12mm or 8-16mm is best.

The use of broken-edged stones reduces conductivity by approximately 50% compared to round gravel, due to turbulent flow within irregular pores. So large grains should be chosen when applying flat or mixed grain shapes, such as chippings from broken stones. In the case of mixed grain size, it is advisable to screen the gravel with the help of a coarse sieve: use the larger grains in the front and the smaller grains in the later sections of the filter. Care must be taken when changing from a larger to a smaller grain size because blockages mostly happen at the point of change.

 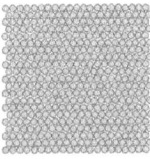 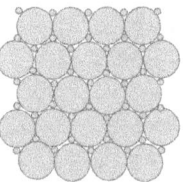 

25mm	5mm	5mm and 25mm	mixed grain size
pore space 22,1%	pore space 45,7%	pore space 23,9%	mixed grain shape
max pore size 2,8	max pore size 0,6	max pore size 1,6	pore space and pore size
spec. surface 143m²/m³	spec. surface 652m²/m³	spec. surface 164m²/m³	unpredictable

Picture 9_19:
Influence of grain size and shape on filter properties

A rather flat slope (α < 45°) should join one grain size to the other to ensure a larger connecting area. In particularly when grain diameters differ considerably, an intermediate zone consisting of intermediate size may be useful. Mixed-grain sizes do not improve hydraulic conductivity. Removing fine soil from gravel by washing is more important than ensuring the exact grain size.

Technical components

If the length of the filter-bed is more than 10m, an intermediate channel for re-distributing cross-flow should be provided. The distribution channel can also serve as a terrace step in the case of steeply sloping topography (Picture 9_21).

The relation between organic load and oxygen supply reduces with length. This happens because oxygen is supplied evenly over the total surface area, whereas the organic load diminishes during treatment. It is most likely, therefore, that anaerobic conditions prevail in the front part, while aerobic conditions reach to a greater depth in the rear part. However, only the upper 5 to 15cm can really be considered an aerobic zone.

A clogged gravel filter can become useful again if it is not used for periods of several months, because of a process called autolysis; when forced to live without feed, the bacteria live on their own bacterial mass.

Figure 9_20
Horizontal gravel filter in India

Filter clogging normally results in surface flow of wastewater. This is usually not desired, although it hardly reduces the treatment efficiency if flow on the surface maintains the assumed retention time inside the filter (this could be the case with dense plant coverage). When filters are well-protected and a long way from residential areas, there is no harm in letting some of the wastewater run above the horizontal surface. Such "overland treatment" produces very good results – especially when the water is equally distributed and does not fester in trenches.

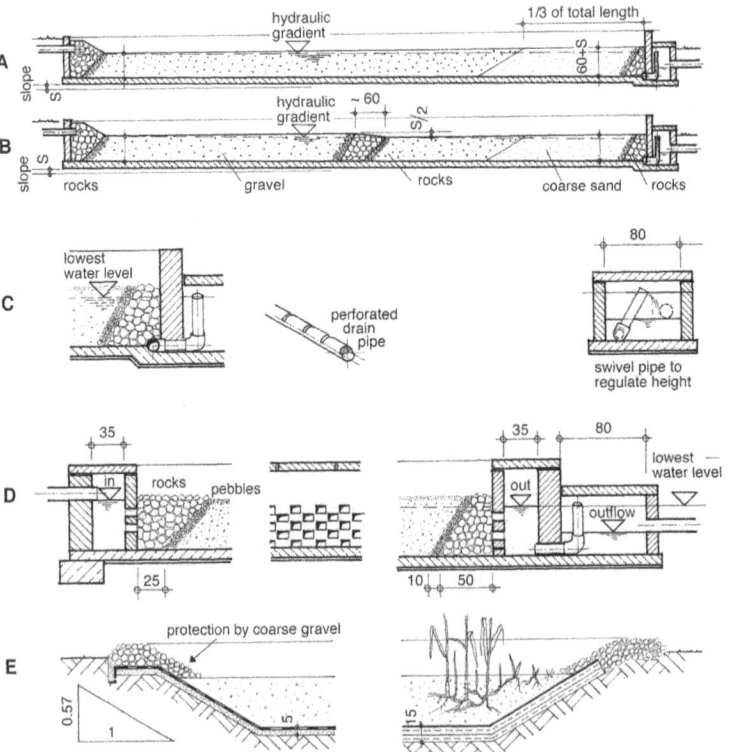

Picture 9_21:
Horizontal gravel filter (subsurface flow filter). A: Filter basin in masonry and concrete structure, finer gravel is used in the rear portion. B: Long filter bed with additional distribution trench in the middle; the trench is filled with rocks and allows a step the surface level. Detail of collection pipe and swivel arm at the outlet side. D: Details of inlet and outlet structure for improved distribution flow for wider filter beds. E: Details of filter basins using foils or clay packs for sealing. Sloped side walls are less costly, but plants will not grow near the rim

Knowledge of the amount of void space within the filter material is essential for calculating the retention time and planning the treatment process. Gravel has 30 to 45% voids, depending on size and shape. (The calculation of HRT in the spreadsheet in table 30 is based on 35% void space; it can be adapted proportional, if the actual void space is greater.) Void space can easily be determined by measuring the water that can be added to a bucket full of gravel (Picture 9_23).

filter medium	diameter of grain mm	pore volume		theoretical conductivity	
		coarse	total	m/s	m/d
gravel	4 - 40	30%	35% - 40%	4.14E - 03	350
sand	0.1 - 4	15%	42%	4.14E - 04	35

Table 20: Theoretical properties of gravel and sand as filter material: lower values should be applied for wastewater when designing filter beds

For high conductivity, large pore size is more important than total pore volume. Pre-wetted gravel shall be used when the pore volume is tested, ensuring that pores of only capillary size are "closed" in advance.

See: Shilton et all, 1996

In reality, short cuts and volume-reduced by partly clogged areas result in 25% shorter retention times and, consequently, inferior performance[34]. For this reason, the filter-bed should not be deeper than the depth to which plant roots can grow (30–60cm), as water will tend to flow faster below the dense cushion of roots. However, treatment performance is generally best in the upper 15cm because of oxygen diffusion from the surface. Shallow filters are more effective, therefore, than deeper beds of the same volume.

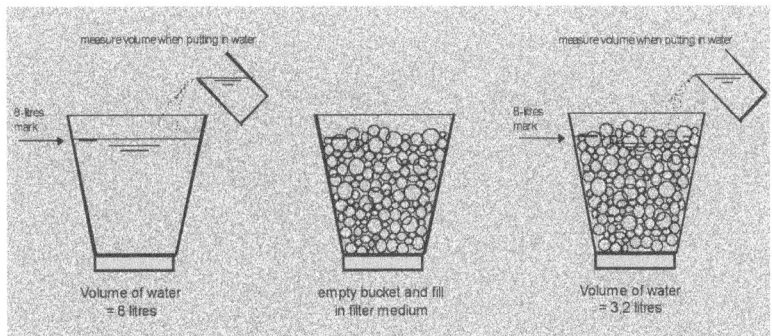

Picture 9_22: Determining pore space of filter medium on site. Example: the empty bucket is full after pouring 8 litre of water. The bucket filled with gravel absorbs 3.2 litre of water: Voids space is 3.2/8 = 0.40 or 40%

Uniform distribution of wastewater throughout the filter requires an equally distributed supply of water at the inlet – and equally distributed reception at the outlet side. Trenches filled with rocks 50 to 100mm in diameter are provided at both ends to serve this purpose. A perforated pipe, which is connected to the outlet pipe, lies below the strip of rocks that form the collection trench. The height of the outlet can be adjusted by a swivel arm, fixed to a flexible elbow. By lifting it until water appears at the surface of the filter near the inlet, the water level in the filter can be adjusted according to hydraulic conductivity. While the top of the filter is kept strictly horizontal to prevent erosion, the bottom slopes down from inlet to outlet ideally at 1%. Site conditions permitting, bigger slope is also possible. To prevent erosion, long filters should have a terraced surface rather than a slope (see picture 9_21 (B)).

The percolation of wastewater into the ground is not desirable so the bottom of the filter must be sealed. While solid-clay packing might be sufficient, heavy plastic foils are more common. A concrete basin with straight, vertical masonry walls allows plants to grow up to the outer rim – not possible with the smooth embankment that plastic foils would require (see picture 9_21 (E)).

In a dry climate, trees search for water and their roots may break the walls and grow into the filter. Whenever possible, trees should not be planted directly beside the filter; this will avoid the structural problems caused by the roots and the unwanted sealing of the filter surface by fallen leaves.

Observation in Europe indicates that the performance of gravel filters diminishes after several years. How long a horizontal filter functions properly depends on several factors: grain size and shape of gravel, the nature and amount of suspended solids in the wastewater, and the temperature and the average loading rate.

If the filter is drained during resting time, alternate charging can increase the treatment performance of horizontal filters. To allow alternate feeding, the total filter area should be divided into several compartments or beds. Other reports recommend that the filter is changed every eight to 15 years. This timeframe is only a rough estimation and – as stated above – depends on the loading rate and structural details, the impact of which is almost impossible to predict in practice. Weaker wastewater, lower loading rates and larger gravel size generally increase the lifetime of the system.

Technical components

Ground filters are covered by suitable plantation – any type of hydro-botanical plant that will grow on wastewater and has deep-reaching and widely spreading roots. The choice of plant influences treatment efficiency; some scientists claim that the micro-environment created inside the filter is responsible for equilibrium between sludge production and sludge "consumption". Such equilibrium is only likely with low loading rates.

Kadlec et all, 1996

The plants are not normally harvested. *Phragmites australis* (reeds), found almost anywhere, are considered to be ideal because their roots form horizontal rhizomes that guarantee a perfect root-zone filter bed (see picture 9_24). Most swamp and water grasses are also suitable, but not all of them have extending or deep-enough roots. Depending on the type of wastewater, different plants might be preferable: *Typha angustifolia* (cat-tails), together with *Scirpus lacustris* (bull rush), have been to be found the most suitable plants for wastewater from petrol refineries, while the large, red- or orange-flowering iris (sometimes known as "mosquito lily") is a beautiful plant, which grows well on wastewater but is only suitable for shallow, domestic gravel beds. Forest trees have also been used and are deemed to be only slightly less efficient[35]. At least two clumps of plants or four sprouted rhizomes should be placed per square metre when planting is started.

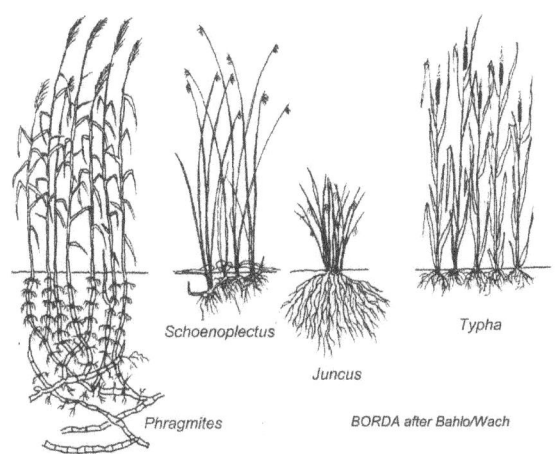

Picture 9_23:
Plant species common for gravelfilter plantation

Within a horizontal filter, plants seem to be "catalysts" rather than "actors" Plants transport oxygen via their roots into the ground. Some scientists claim that this process also supplies surplus oxygen, thereby creating an aerobic environment, while others have shown that plants only transfer as much oxygen as they need to fulfil their own nutrient requirements. For example, Brix and Schierup claim that plants provide 0.02g $O_2/m^2 \times d$ to the filter bed, while consuming 2.06g $O_2/m^2 \times d$ for themselves. Nonetheless, it is assumed that toxic substances near the roots are eliminated by oxidation. The complex ecosystem that exists in planted gravel filters produces good and reliable treatment results, which in part, must be due to aerobic treatment. This is underlined by reports which claim that COD reduction rates of over 95% can be achieved – which would not be possible under anaerobic conditions alone. The uptake of nutrients by plants is of relatively little importance, especially when plants are not harvested.

Starting phase and maintenance

Young plant seedlings may not grow on wastewater. So it is advisable to start feeding the plant with plenty of fresh water and to let the pollution load grow parallel to plant growth.

When plants are under full load, the outlet level is adjusted according to flow. Water should not stand on the surface near the inlet. If this happens, the swivel arm at the outlet should be lowered. Optimal water distribution at the inlet side is important and must be controlled from time to time. Replacement of the filter media might be necessary when treatment efficiency declines. Since there is no treatment during the time that the filter media is being replaced, it is advantageous to install several, parallel filter-beds.

To prevent clogging of the filter with fine soil, stormwater should neither be mixed with the wastewater before the treatment step, nor should outside stormwater be allowed to overflow the filter bed. Erosion trenches around the filter-bed should always be kept in proper functioning condition.

Technical components

Calculating dimensions

If percolation properties – the so called hydraulic conductivity of the filter body – is known, then the required cross-sectional area at the inlet can be calculated using Darcy's Law. To compensate for reduced conductivity with use, only a fraction of the calculated figure for clear water should be used for designing the plant. The conductivity applied in the spreadsheet takes this into consideration. It does not, however, take head of pessimistic statements, which claim that only 4% of the clear-water conductivity should be used. The dimensions of the filter depend on hydraulic and organic loading, temperature and grain size of the filter medium. As a rule of thumb, 5m² of filter should be provided per capita for domestic wastewater. This would mean a hydraulic loading rate of 30l/m² and an organic loading rate of 8g BOD/m² x d. For comprehensive calculation use the formula applied in the computer spread sheet (Table 30, page 264).

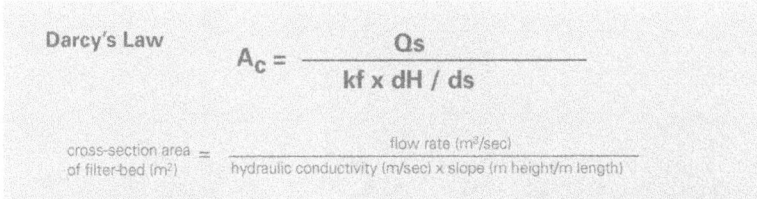

Darcy's Law

$$A_c = \frac{Q_s}{k_f \times dH/ds}$$

cross-section area of filter-bed (m²) = flow rate (m³/sec) / hydraulic conductivity (m/sec) x slope (m height/m length)

Figure 9_24:
Darcy's law for calculation of hydraulic conductivity

Figure 9_25:
Horizontal gravel filter during construction. Constructed above ground

9.2.7.2 Vertical sand filter

Although the vertical filter is – compared to the horizontal filter – the more efficient and more reliable treatment system from a technical and scientific point of view, it is not suitable for DEWATS because of its permanent operational control, necessity of a dosing device, and strict adherence to charging intervals. Nonetheless, the following section introduces this system to provide a better understanding of related treatment processes.

The vertical filter functions in a similar way to an aerobic trickling filter and, consequently, must be fed at intervals with defined resting times between dosing charges. In addition to the short intervals, which are regulated by dosing devices, longer resting periods of one to two weeks are required. This is only possible if there are at least two alternately fed filter beds.

Feeding in doses is necessary for equal water distribution. The resting times are needed so that oxygen can enter the filter after wastewater has percolated (see picture 9_17 on page 196). Doses must be large enough to temporarily flood the complete filter and to distribute the water evenly over the surface, but small enough to allow enough time for oxygen to enter before the next flooding. The filter material, therefore, must be fine enough to cause flooding and porous enough to allow quick percolation. During the short charging times, the wastewater is exposed to the open, which can create a bad odour in the case of anaerobic pre-treatment.

The body of the vertical filter consist of a fine top layer, a medium middle layer and a rough bottom layer. The area below the filter media is a free-flow area, connected to a drainpipe. The free-flow area is also connected to the open via additional vent pipes. The fine top layer guarantees homogeneous flow distribution; the middle layer is the actual treatment zone, while the bottom layer is responsible for providing wide-open pores to reduce the capillary forces, which would otherwise decrease the effective hydraulic gradient.

Technical components

Vertical filters are normally 1m to 1.20m deep. However, if there is enough natural slope and good ventilation, vertical filters can be constructed up to three metres high. Vertical filters may or may not be covered by plantation. In the absence of plantation, the surface must be scratched at the beginning of each resting period, in order to allow enough oxygen to enter; with dense plantation, the stems of the plants ensure sufficient open pores in the filter surface. Several charging points are distributed over the surface to allow quick flooding of the full area. Flooding is the only reliable method of achieving equal distribution of water over the entire filter; charging points spaced across the surface area allow quick submergence. It is not possible to achieve equal distribution by designing supply pipes of different diameters and length, leading to various outlet points. This has been tried often enough; we don't need new failures. Flush distribution is a must.

Dosing of flow can be regulated with self-acting siphons, automatic controlled pumps or tipping-buckets. The latter is most suitable under DEWATS conditions because its dertermining principle is easily understood and the hardware can be manufactured locally.

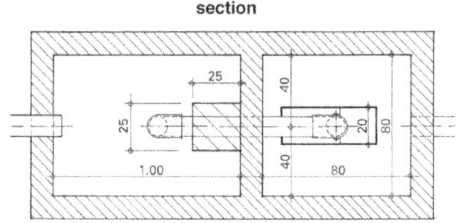
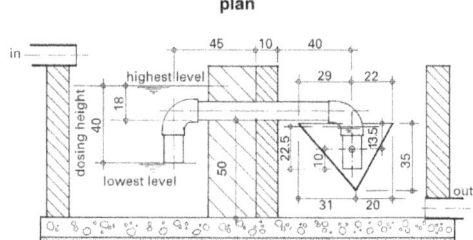

Picture 9_26:
Dosing chamber with tipping bucket for the controlled operation of a siphon. The bucket closes the siphon until it is filled with water. When losing its equilibrium due to the weight of the water, the bucket turns over and opens the siphon. It falls back into horizontal position to receive new water, which again closes the siphon for the next flush

The flow to each bed can be prevented or controlled when necessary with a valve within the inlet pipe. Alternatively, the valve can be replaced by a straight standing piece of pipe in the dosing chamber (see Picture 9_28).

While vertical filters can bear a hydraulic load up to $100 l/m^2 \times d$ ($100 mm/m^2 = 0.1 m$), it is better to restrict loading to $50 l/m^2 \times d$. The organic load may reach up to $20 g\ BOD/m^3 \times d$; in the case of re-circulation, $40 g\ BOD/m^2 \times d$ is possible (Metcalf & Eddy). In the case of pre-treated domestic wastewater, the hydraulic load is the deciding factor. Some engineers use these values only for active filter-beds, while others claim that the resting beds must be included within the calculation. If there is any doubt, testing is recommended. However, larger filter areas are always preferable.

Permeability can be calculated with Darcy's Law (on page 206), whereas $dH/ds = 1$. The flow speed ($v = Qs/Ac$), therefore, is equal to the hydraulic conductivity (k).

Starting phase, maintenance and calculating dimensions

The vertical sand filter does not belong to DEWATS. Detailed operational instructions have been deliberately excluded from this handbook to ensure readers don't get the impression that the vertical filter can be constructed and operated under DEWATS conditions.

Technical components

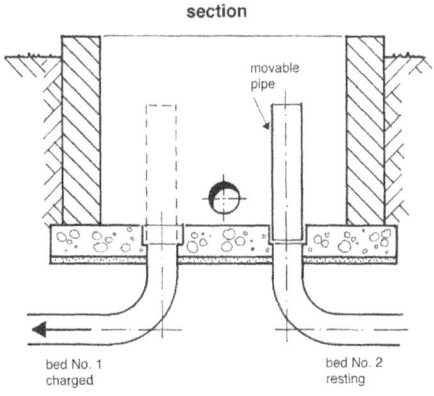

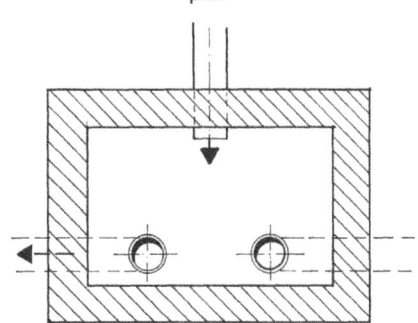

Figure 9_27:
Distribution chamber for alternate feeding of filter beds. A piece of straight pipe is placed on the outlet, which is to be temporarily closed

9.2.8 Ponds

Ponds (lagoons) are artificial lakes. They provide wastewater treatment through natural processes. Different treatment processes can be utilised; depending on the design of the artificial lake, series of ponds can be used to combine different treatment effects. Ponds are ideal DEWATS and should be given preference over other systems whenever land is available. Ponds are preferable to underground gravel filters, if sympathetic to the surroundings; facultative or anaerobic ponds must be far enough from human settlements to avoid the nuisance caused by bad odours or mosquito breeding. Polishing ponds can be closer, if fish are held within the water body; fish that belong to *gambusia spp.* are commonly used for mosquito control in tropical countries.

Pure pond systems are cheap and need almost no maintenance, even if large.

Ponds may be classified into:
- sedimentation ponds (pre-treatment ponds with anaerobic sludge stabilisation)
- anaerobic ponds (anaerobic stabilisation ponds)
- oxidation ponds (aerobic cum facultative stabilisation ponds)
- polishing ponds (fully-aerobic post-treatment ponds, placed after stabilisation ponds)

Pond systems intended to provide full treatment normally consist of several ponds serving different purposes. For example, a deep anaerobic sedimentation pond for sedimentation cum anaerobic stabilisation of sludge, two or three shallow aerobic and facultative oxidation ponds with longer retention times for predominantly aerobic degradation of suspended and dissolved matter, and one or several shallow polishing ponds for the final sedimentation of suspended stabilised solids and bacterial mass. Wastewater ponds for fish farming require low organic loading and, in addition, should be diluted by four to five times the amount of river water. Otherwise, the pond must be about 10 times as large as the area calculated in the spreadsheet (see Table 33, page 273).

Artificially aerated ponds are not considered to be DEWATS and, therefore, are not dealt with in this handbook. It may be enough to know that such ponds are 1.5 to 3.5m deep, usually work with a five days hydraulic retention time (HRT) and organic loads of 20 to 30g $BOD/m^3 \times d$. The energy requirement for aeration is about $1-3W/m^3$ of pond volume. Only where there is a little scum only the surface of anaerobic ponds may be aerated to reduce the foul smell.

9.2.8.1 Anaerobic ponds

Anaerobic ponds are deep (2 to 6m) and highly loaded (0.1 to 1kg BOD/m³×d). Anaerobic conditions are guaranteed by the depth of the pond, thereby requiring less surface area than aerobic-facultative oxidation ponds.

It is possible to provide separate sludge-settling tanks before the main pond, in order to reduce the organic-sludge load. Such settling tanks should have a HRT of less than one day, with the exact HRT depending on the kind of wastewater. Anaerobic ponds with an organic loading rate of below 300g/m³×d BOD are likely to remain at an almost neutral pH. Consequently, they release little H_2S and, therefore, are almost free from an unpleasant smell. Highly loaded anaerobic ponds omit foul odour, until a heavy layer of scum has been developed.

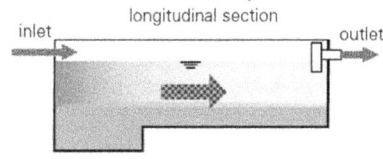

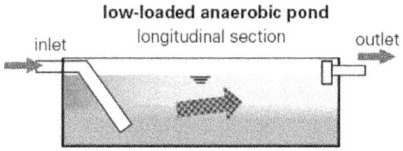

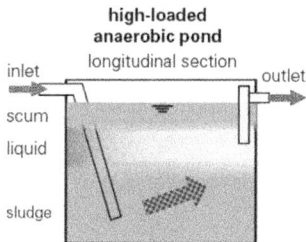

Figure 9_28: Principles of anaerobic ponds. Sedimentation ponds have a HRT of about one day, ponds with low loading are supposed to be odourless because of almost neutral pH, highly loaded ponds form sealing scum layer on top

Before this layer exists, the upper region of the pond will remain aerobic; these ponds are called facultative-anaerobic.

Depending on the properties of the wastewater, the desired treatment effect and possible post-treatment, anaerobic ponds are designed for hydraulic retention times of between one and 30 days. The HRT determines whether only settled sludge or all of the liquid is treated. For domestic wastewater the anaerobic pond may function as an open septic tank. It should be small, in order to develop a sealing scum layer; in this case, treatment efficiency is only in the range of 50 to 70% BOD removal.

"Wrong" retention times result in stinky effluent. If the retention time is longer than one day, not only bottom sludge but also the liquid portion begins to ferment. On the other hand, if the retention time is too short for the liquid to stabilise substantially, the effluent remains at a low pH and stinks of H_2S. Too-short retention times have the same effect as too-high organic loading rates.

pollutant	dimension	inflow	outflow	removal rate
suspended solids	mg/l	431	139	68%
COD mg/l	mg/l	1189*	505	58%
BOD_5	mg/l	374	190	49%
Nkjel	mg N/l	116	99	15%
P total	mg/l	26	24.5	6%
fecal coli	No/100ml	6,156,000	496,000	92%
fecal strepto	No/100ml	20,900,000	1,603,000	92%
nematode ova	No	139	32	77%
cestode ova	No	75	18	76%
helminth ova	No	214	47	78%

*the high COD/BOD ratio is caused by mineral of pollution which is also the reason for the COD-removal rate being higher than that of the BOD_5

Table 21:
An example of high performan of a simple sett pond
Source: Drioach all, 1997

Technical components

In industries, such as sugar plants or distilleries, anaerobic ponds are often used as the first treatment unit, followed by oxidation ponds. The treatment efficiency of high-loaded ponds with long retention times ranges from 70 to 95% BOD removal (CODrem. 65 to 90%), depending on the biodegradability of the wastewater. Several ponds in series are recommended for long retention times.

Anaerobic ponds are not very efficient in treating wastewater with a wide COD/BOD ratio (> 3:1). For this type of wastewater, sedimentation ponds with very short retention times, followed by aerobic/facultative stabilisation ponds are recommended.

ambient temperature °C	org. load BOD g/m²*d	efficiency BOD rem. %
10	100	40
15	200	50
20	300	60
23	330	66
25	350	70
28	380	70
30	400	70
33	430	70

Table 22:
Design parameters for low-loaded anaerobic ponds in relation to ambient temperature
Source: Mara 1997

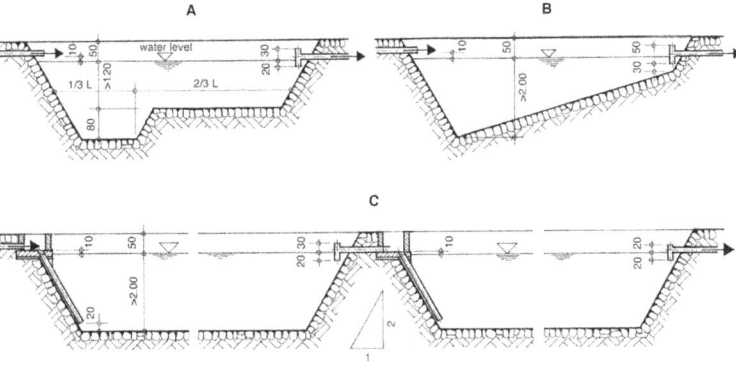

Picture 9_29:
Cross-sections of anaerobic ponds constructed from rocks, with cement-mortar pointing.
A and B: A deeper inlet section accumulates most of the sludge within a limited surface area.
C: Two anaerobic ponds in series. The first pond may be highly loaded (scum sealed), while the second pond may be low loaded (neutral pH)

Pond size is also determined by the long-term sludge-storage volume. Anaerobic ponds with sufficient, integrated sludge storage make sludge-removal intervals of over 10 years possible.

Starting phase and maintenance

Start-up does not require any special arrangements. But one must be aware of the fact that a heavily loaded pond will release bad odour until a layer of scum seals the surface. Inlet and outlet structures should be monitored during operation. A drop in the effluent quality is a warning that the sludge must be removed. If this is neglected, the receiving waters or the ensuing treatment units will suffer the consequences.

Calculating dimensions

Retention time and volumetric organic load are the two design parameters for anaerobic ponds. A non-smelling pond loaded with 300g $BOD/m^3 \times d$, for a short HRT of one day, requires approximately $0.2m^3$ per capita for domestic wastewater. Anaerobic stabilisation of the liquid fraction requires longer retention times, the calculation of which depends on temperature, desired treatment quality and organic load. The organic loading rate should not exceed 1kg $BOD/m^3 \times d$. For exact calculation please refer to the formula in the spreadsheet (Table 31 and 32, pages 268 and 269).

Picture 9_30: Anaerobic pond in Namibia

9.2.8.2 Aerobic ponds

Aerobic ponds receive most of their oxygen via the water surface. For loading rates below 4g BOD/m²×d, surface oxygen can meet the full oxygen demand. Oxygen intake increases at lower temperatures and with surface turbulence caused by wind and rain. Oxygen intake also depends on the actual oxygen deficit up to saturation point so may vary at 20°C between 40g O_2 /m²×d for fully anaerobic conditions and 10g O_2 /m²×d in the case of 75% oxygen saturation.[36] (Mudrak&Kunst, after Ottmann 1977).

See Mudrake et all, 1997

The secondary source of oxygen comes from algae via photosynthesis. However, in general, overly intensive growth of algae and highly turbid water prevents sunlight from reaching the lower strata of the pond. Oxygen "production" is then reduced because photosynthesis cannot take place. The result is a foul smell because anaerobic facultative conditions prevail. Algae are important and positive for the treatment process, but are a negative factor when it comes to effluent quality. Consequently, algae growth is allowed and wanted in the beginning of treatment, but not desired when it comes to the point of discharge because algae increase the BOD of the effluent. Algae in the effluent can be reduced by a small final pond with a maximum one-day retention time. Larger pond areas – low loading rates with reduced nutrient supply for algae - are the most secure, but also the most expensive measure.

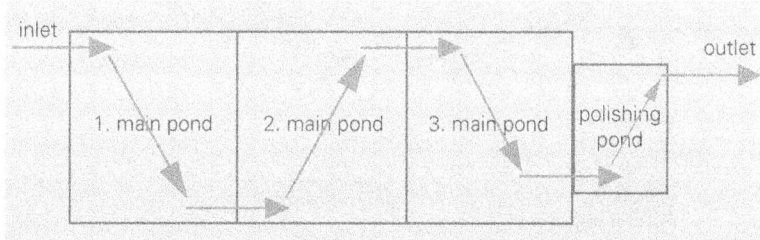

ture 9_31:
Flow pattern of aerobic-facultative ponds in series

The laboratory results of effluent wastewater often give a false impression of insufficient treatment. As nearly 90% of the effluent BOD comes from algae, many countries allow higher BOD loads in the effluent from ponds, as compared to other treatment systems. Baffles or rock bedding before the outlet of each of the ponds have a remarkable effect of algae retention. Intelligent structural details increase the treatment quality considerably at hardly any additional cost – and may be seen as being as important as adequate pond size.

Treatment efficiency increases with longer retention times. The number of ponds is of only relative influence. With the same total surface, splitting one pond into two ponds increases efficiency by approximately 10%. Having three instead of two ponds adds about 4% and from three to four ponds having inreases efficiency by another 2%.

This shows that having more than three ponds is not justifiable from an economic point of view because the same effect can be achieved by just enlarging the surface area. Instead of constructing the dams and banks of an additional pond, the required land should be used as additional water area.

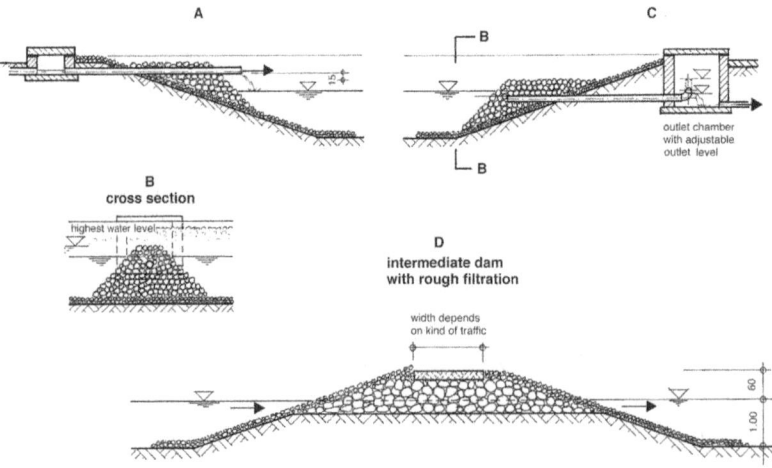

Picture 9_32:
Section through a large aerobic-facultative stabili sation pond. Ban should be protec against erosion by waves. A: Inle banks should also be protected against erosion t influent. B: Cross section B-B (from view of C). C: Outlet structure with swivel arm to adjust height c pond according t seasonal fluctuat of water volume.

Technical components

The first pond may be as much as twice the size of the others, if there are several inlet points. In principle, having several inlet points – to distribute the pollution load more equally and to create a larger area for sedimentation – is an advantage. On the other hand, it might be advisable to provide a slightly separated inlet zone to avoid bulky, floating matter littering the total pond surface.

The inlet points should be as far away from the outlet as possible. The outlet should be below the water surface to retain floating solids, including algae. Gravel beds acting as roughing filters are advisable between ponds in series and before the final outlet.

The erosion of banks by waves could be a problem with larger ponds. Therefore, the slope should be 1 (vertical) to 3 (horizontal) and covered with rocks or large bits of gravel. This also helps to keep the soil embankments from slipping. Banks and dams can be protected by planting macrophytes, such as cat-tail, or phragmites. Dams between ponds should be paved and wide enough to facilitate maintenance.

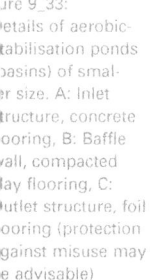

Figure 9_33: Details of aerobic stabilisation ponds (basins) of smaller size. A: Inlet structure, concrete flooring, B: Baffle wall, compacted clay flooring, C: Outlet structure, foil flooring (protection against misuse may be advisable)

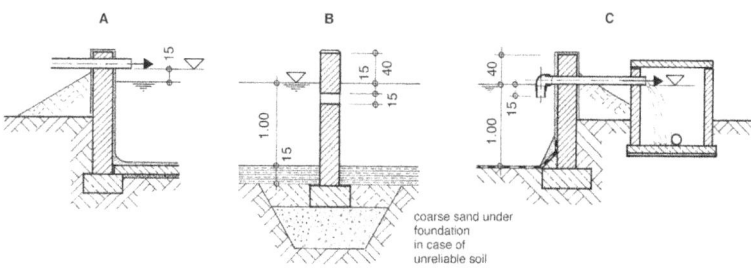

Aerobic stabilisation ponds must be shallow enough to permit adequate oxygen intake but deep enough to prevent weed growth at the bottom of the pond. A depth of 90cm to 1m in a warm climate and up to 1.2m in cold-climate zones (due to frost) is suitable. Deeper ponds become facultative or even anaerobic in the lower strata.

Smaller volumes of wastewater, such as from schools, hospitals or residential houses are better pre-treated in Imhoff tanks, septic tanks, baffled reactors or, at the least, sedimentation pits, before reaching the aerobic stabilisation pond. Properly operated Imhoff tanks, which have odourless effluent are preferable. A septic tank with smelly effluent is to be preferred if regular desludging of the Imhoff tank cannot be guaranteed. If pre-treatment is not provided, the pond must have a deeper sedimentation zone near the inlet; bad odour is to be expected. It might be wiser, therefore, to construct a small sedimentation pond, on which a sealing scum layer will develop. Should the scum layer reach a thickness of more than 10cm, papyrus can be grown on it to make it look more attractive.

Starting phase and maintenance

The pond matures much faster if it is filled with river water *before* the first wastewater enters. With the exception of controlling the inlet and outlet structures regulary, no permanent attendance is required. But the performance of the pond should be monitored and any disturbance of the water quality investigated. Sludge must be removed at defined intervals, to avoid a decline in treatment quality. Adequate sludge removal, handling and treatment is discussed in detail in section 11.3.

Technical components

Calculating dimensions

Organic surface load and hydraulic retention time are the two decisive design parameters. While the minimum hydraulic retention time ranges from five to 20 days, the maximum organic load depends on the ambient temperature (see Table 23 on this page). The amount of sunshine hours is important, as UV radiation is effective at destroying pathogens. Although this consideration is not included in the calculation, ponds should be slightly oversized in areas with permanent cloud cover. Organic loading should be less than 20g BOD/m²×d. For domestic wastewater, a pond surface of between 2.5 and 10m² may be estimated per capita. All values depend on the type of pre-treatment, the surrounding temperature, and health objectives. For more exact calculation, please refer to the formula applied in the computer spreadsheet (see Table 33, page 273).

Table 23:
Organic surface loading for aerobic-facultative ponds
Source: Mara, 1997

5 days HRT	
ambient temperature °C	organic load BOD g/m²*d
10	7.0
15	11.7
20	17.7
23	21.8
25	24.5
28	28.4
30	30.8
33	33.8

9.2.9 Hybrid and combined systems

Each technology has its strengths and weaknesses. It makes sense, therefore, to combine different treatment units into a more efficient modular, treatment system. An example of such a combined system could be sedimentation in a settler or septic tank followed by anaerobic decomposition of non-settleable suspended solids in anaerobic filters or baffled reactors. Further treatment in ponds or ground filters provides aerobic conditions. Deciding which technologies are most appropriate for combining depends on treatment requirements and boundary-site conditions.

Picture 9_34: The complete treatment chain of DEWATS-technology

Apart from applying different DEWATS modules in series, hybrid systems can also combine different technologies within one treatment unit. One can, for example, combine the anaerobic baffled reactor with the anaerobic filter by adding filters in the last chambers (see picture 9_36); alternatively, if a floating-filter medium is available, one may provide a thin filter layer at the top of each baffled chamber. In practice, the combination of six baffled chambers with two filter chambers has performed reliably and well.

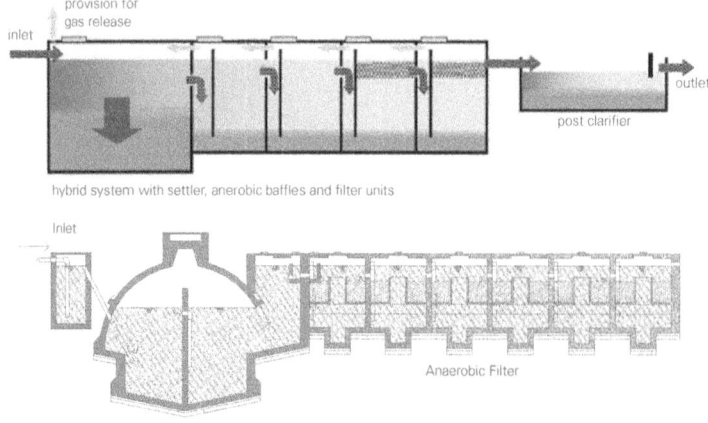

Picture 9_35: Examples of a hybrid and a combined system

Technical components

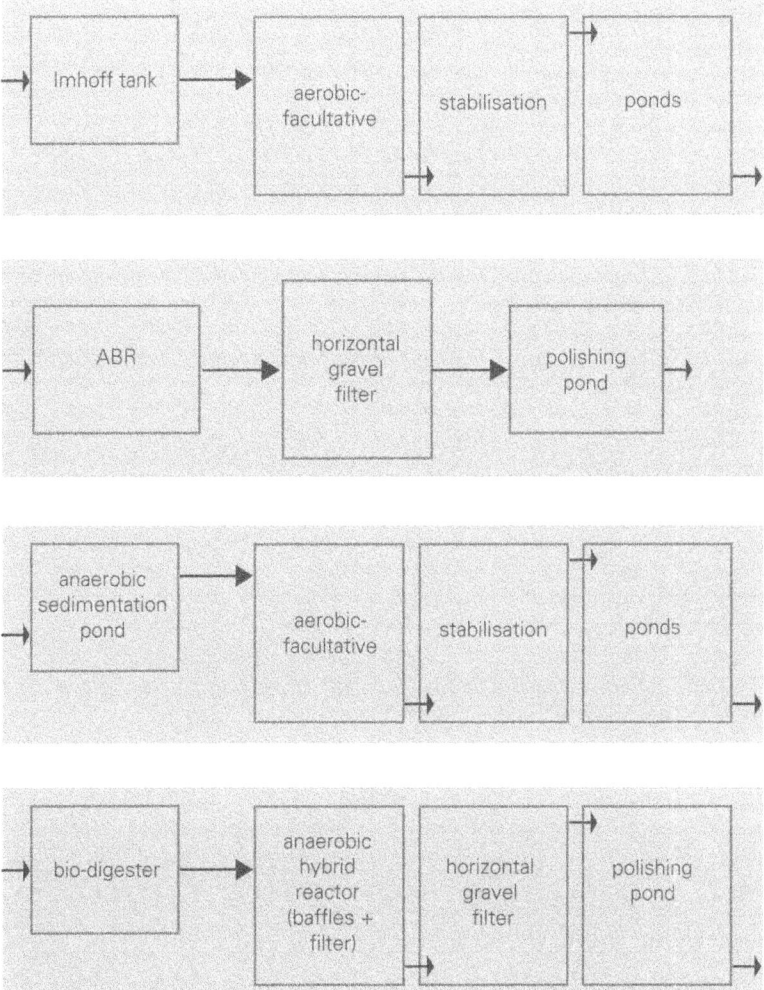

Figure 9_36: Typical combinations for full treatment with DEWATS-modules

9.3 Non-DEWATS technologies

DEWATS commands low maintenance. This implies that technologies which cannot be "switched on and off" as one likes, is integral to the DEWATS concept. DEWATS are intended to function every day with the efficiency envisaged. Systems, which are highly efficient but require a great deal of regular care to function at an acceptable level, do not suit the concept of decentralised wastewater treatment. To avoid any misunderstanding: The technologies which are regarded here as non-DEWATS are by no means inferior treatment systems. They may even be used in a decentralised concept. However, not without highly qualified operational staff which is closely supervised by an experienced management.

9.3.1 UASB

The UASB system is not considered a DEWATS technology. However, an understanding of the principle on which it functions may improve one's understanding of the anaerobic baffled reactor.

In a UASB reactor (Upstream Anaerobic Sludge Blanket reactor) the upstream velocity and settling speed of the sludge is in equilibrium and forms a locally rather stable, suspended sludge blanket in the lower part of the digester. This sludge blanket of suspended active sludge acts as a filter medium. After some weeks of maturation, granular sludge forms and improves the physical stability and filter capacity of the sludge blanket.

To keep the blanket in its proper position, the hydraulic load must correspond with the upstream velocity and with the organic load. The latter is responsible for the development of new sludge. So the flow rate must be controlled and regulated in accordance with fluctuations of the organic load. Generally, the fluctuation of inflow is high in smaller units and regulating wastewater flow is not possible. Furthermore, it is not possible to stabilise the process by increasing the hydraulic retention time without lowering the upstream velocity. Although the system is simple to build, these operational difficulties render it unsuitable for DEWATS, particularly for relatively weak, domestic wastewater.

Technical components

Fully controlled UASBs are used for relatively strong industrial wastewater. Slanting baffles (similar to the Imhoff tank) help to separate gas bubbles from solids, whereby solids are also separated from the up-streaming liquid. These baffles are called 3-phase separators. Biogas can be collected and used.

UASB reactors require several months to mature – to develop sufficient granular sludge to provide treatment. Granular sludge looks like big flocs of dust; microbial slime forms chains, which coagulate into flocs or granules. High organic loading, in connection with lower hydraulic loading, speeds up the granulation process in the starting phase. Since higher velocities are required to lift sludge granules compared to single sludge particles, the sludge blanket remains relatively stable.

Starting phase, maintenance and calculating dimensions

The UASB does not belong to DEWATS. Details of how to operate it and calculate its dimensions are deliberately omitted from this handbook so that readers don't gain the impression that the UASB can be built and operated under DEWATS conditions.

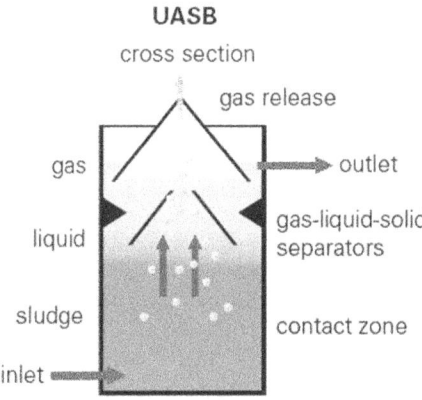

Figure 9_37:
Flow principle of UASB reactors. Up-streaming water and gas-driven sludge particles hit baffles, which cause to separate gas, solids and liquid

9.3.2 Trickling filter

The trickling filter is not thought of as a DEWATS solution. But some understanding of how it works will improve one's understanding of the principle of aerobic-wastewater treatment.

The trickling filter follows the same principle as the anaerobic filter, in the sense that it provides a large surface for bacteria to settle. The main difference between the two systems lies in the fact that the trickling filter works under aerobic conditions. This implies that the bacteria, which are immobilised at the filter medium, must have equal access to air and wastewater. So wastewater is dosed at intervals, providing time for air to enter the reactor during the breaks. An equal distribution of wastewater over the full surface area will utilise the filter mass most efficiently.

A trickling filter consists of:
- a dosing device
- a rotating sprinkler
- a filter body, which is ventilated both from the top and the bottom

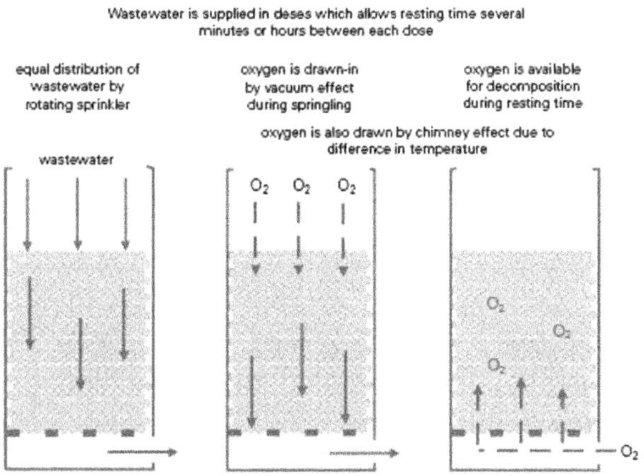

Technical components

Wastewater is supplied in doses which allows resting time several minutes or hours between each dose

- equal distribution of wastewater by rotating sprinkler
- oxygen is drawn-in by vacuum effect during springling
- oxygen is available for decomposition during resting time
- oxygen is also drawn by chimney effect due to difference in temperature

Picture 9_38:
The principle of the trickling filter

Rocks with a diameter of between 3 and 8cm in are used as the filter medium. The outside of the filter body is closed to prevent sludge flies from escaping into the open. The filter rests above ground to allow ventilation. The bottom slab is sloped so that sludge and water inses away. The bacterial film must be flushed away regularly to remove dead sludge and to prevent clogging. High hydraulic-loading rates (> $0.8 m^3/m^2 \times h$) have a self-flushing effect. With organic-loading rates of 1kg $BOD/m^3 \times d$, 80% BOD removal is possible. Higher loading rates reduce efficiency.

In a 2m-high trickling filter with a wastewater of 500mgBOD/l, the organic-loading rate comes to:

$$0.8 \times 24h \times 0.5 kgBOD/m^3 / 2m \text{ height} = 4.8 kg\ BOD/m^3 \times d$$

At such a high organic load, a removal rate of only 60% BOD may be expected. The simple calculation shows that wastewater would have to be recycled nearly five times to get the expected treatment quality and self-flushing effect. However, the trickling filter could be operated with lower hydraulic-loading rates if regular flushing is done.

The self-flushing (high-rate) trickling filter is a reliable system, despite fluctuations in the flow of wastewater. Nonetheless – as it requires a rotating sprinkler and pump for operation – the system is not a suitable DEWATS solution.

Starting phase, maintenance and calculating of dimensions

Details for calculation and instructions for operation are not included in this handbook as the trickling filter cannot be built and operated under DEWATS conditions.

9.3.3 Aquatic-plant systems

Water hyacinth, duckweed, water cabbage and other aquatic plants can improve the treatment capacity of pond systems. The heavy metals that accumulate in water hyacinths are removed when the plants are harvested. Duckweed is a good substitute for algae; if not confined within fixed frames, duckweed is blown by the wind to the lee-side of the pond. If it is retained in a surface baffle, it leaves a cleaner effluent. Improved treatment efficiency however, is only guaranteed by regular attendance and harvesting. Special design features for harvesting increase the total area requirement of the treatment system. The evaporation rate of aquatic-plant systems is four times higher than that of open ponds (in the range of 40l/m²×d in hot climates).

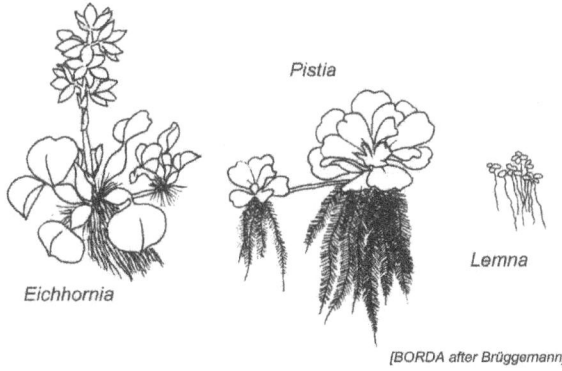

Picture 9_39: Aquatic plants, commonly used for wastewater treatment

[BORDA after Brüggemann]

The area required for a pond is almost the same, regardless of aquatic plants. If the organic-loading rate is low, plants provide protect mosquito-controlling fish from birds. However, some plants such as water hyacinth, are disadvantageous, as they hide mosquito larva from fish and provide shelter for snakes. High organic loading rates – where additional treatment by aquatic plants is most beneficial – do not allow the survival of fish for mosquito control.

As aquatic-plant systems become a nuisance if they are not maintained properly, they are not considered as DEWATS. However, aquatic plants make sense if utilised in conjunction with wastewater farming for intensive and controlled nutrient recycling, or to improve the appearance of residential areas.

Starting phase and maintenance

Operation and maintenance is mainly an agricultural-management issue rather than a wastewater-treatment issue. The pond should start off with fresh, river water and the pollution load should be slowly increased, as plant cover increases. Plants must be harvested regularly to prevent bottom sludge forming from dead plants. Duckweed, in particular, should be kept within frames. Inlet and outlet structures should be controlled regularly.

Calculating dimensions

For practical reasons, please refer to the same formula as for unplanted oxidation ponds (see Table 33, page 273).

Picture 9_40:
Aerobic pond for post-treatment of a DEWATS system at Aravind-Eye-Hospital in Pondicherry, India

Designing DEWATS

If the planning engineer knows his or her craft and recognises his or her limitations, designing DEWATS is relatively simple. Treatment-system performance cannot be precisely predicted and, therefore, calculating of dimensions should not involve ambitious procedures; in the case of small- and medium-scale DEWATS, a slightly oversized plant volume adds to operational safety.

Based on local conditions, needs and preferences, plants of varying sizes can be chosen as standard designs. On-site adaptations can then be made by less-qualified site supervisors or technicians.

In the case of specific demands, calculations and design must be carried out individually; the structural details of the standardised plants can be integrated. In this chapter we introduce a simplified, quasi-standardised method of calculating dimensions using spreadsheets.

Co-operative plant systems that require interconnecting sewerage must be designed individually by an experienced engineer, who is able to place plants and sewers according to contours and other site requirements.

10.1 Technical spreadsheets – background

10.1.1 Usefulness of computer calculation

The purpose of this chapter is to provide the engineer with tools to produce his or her own spreadsheets for sizing DEWATS in any computer programme that he or she is familiar with. The exercise of producing one's own tables will compel engineers to deepen their understanding of design.

The curves that have been used as the basis for calculation in the formulas applied in the computer spreadsheets may also be of interest to those who do not use a computer (these are found in this chapter). As these curves visualise the most important relationships between various parameters, they will enhance understanding of the factors that influence the treatment process. It should be noticed that the graphs have been developed on the basis of mixed information; the methods of calculation, therefore, do not always follow the same logic.

Computerised calculations can be very helpful, particularly if the formulas and the input data are correct. Flawed assumptions or wrong data, on the other hand, will definitely result in worthless results. Nevertheless, assuming the input data is correct, spreadsheets provide a quick impression of the plant's space requirement and what treatment performance can be expected. Ready-to-use computer spreadsheets are especially helpful to those who do not design DEWATS on a daily basis and would otherwise need to recollect the entire theory for sizing a plant before starting to design.

Please bear in mind that DEWATS provides a set of approaches. The equations used in the technical spreadsheets do rely on certain assumptions. Because of the very different parameters that are relevant for the performances of a plan (temperature, materials to be used, composition of the wastewater etc.) there is not a "right way" to calculate dimensions. It is the experience and understanding of the planner that is crucial to create the designs most appropriate to local conditions – i. e. the wastewater problem.

10.1.2 Risks of using simplified formulas

The formulas applied in the spreadsheets have been developed by practitioners, who are not overly concerned with theoretical knowledge. But the formulas are based on scientific findings, which have been simplified in the light of of practical experience.

Even if the formulas were to be 100% correct, the results would not be 100% accurate, as input data is not fully reliable. But the accuracy of the formulas is likely to be greater than the accuracy of wastewater sampling and analysis. There are many unknown factors influencing treatment efficiency and "scientific" handbooks provide a possible range of results. But this book, although "scientifically" based, is written for people who have to build a real plant out of real building materials. The supervisor cannot tell the mason to make a concrete tank "about 4.90m to 5.60m long"; he or she must say: "The length should be 5.35m". The following spreadsheets were designed in this spirit. Anyone who already uses more variable methods of calculation and who is not the target reader of this book is free to modify the formulas and curves according to his or her experience and ability (the authors welcome any information that would help to improve the spreadsheets).

As the formulas represent simplifications of complex natural processes, there is a certain risk that they do not reflect reality adequately. However, the risk of changes in the assumed reality is even greater; for example, expanding a factory without enlarging the treatment system is obviously more significant than an assumed BOD of 350mg/l, when in reality it is only 300mg/l.

Listed below are some examples of incorrect assumptions and their consequences:
- underestimating sludge accumulation in septic tanks, sedimentation ponds, Imhoff tanks and anaerobic reactors results in shorter desludging intervals
- in the case of anaerobic reactors, severe under-sizing could lead to a collapse of the process, while over-sizing may require longer maturation time at the beginning
- incorrect treatment performance of primary or secondary treatment steps could be the cause of over- or undersized post-treatment facilities. This may result in unnecessarily high investment costs or having to enlarge the post-treatment facilities
- undersized anaerobic ponds will develop odour, while slightly oversized ponds may not develop sufficient scum, also resulting in smells
- undersized aerobic ponds can develop an odour; there is no harm in oversizing aerobic ponds
- the biggest risk lies in filter media clogging in both anaerobic tanks and constructed wetlands. However, the risk is more likely to come from inferior filter material, faulty structural details or incorrect wastewater data than from incorrect sizing

In general, moderate oversizing reduces the risk of unstable processes and inferior treatment results.

10.1.3 About the spreadsheets

The spreadsheets presented in this handbook are in Microsoft EXCEL; other suitable programmes may also be used.

There might be differences in the syntax of formulas, for example 3^2 (3 to the power of 2) may be written as =POWER(3;2) or =3^2, square root of 9 could be =SQRT(9) or =9^1/2, cubic root of 27 would be =power(27;1/3) or =27^1/3. Some programmes may accept only one of the alternatives.

The spreadsheets are based on data which is normally available to the planning engineer within the context of DEWATS. For example, while the measurement of BOD_5 and COD may be possible at the beginning of planning, it is unlikely that the BOD_5 will be regularly controlled later on. Therefore, calculations are based on COD or the results of BOD-based formulas have been set in relation to COD, and vice versa. In the following, the term BOD stands for BOD_5.

The formulas applied in the spreadsheets are based on curves from scientific publications, handbooks and the experience of BORDA and its partners. The formulas, therefore, define typical trends. For example, it is well-known that the removal efficiency of an anaerobic reactor increases when the COD/BOD ratio is narrow. Such curves have been simplified into a chain of straight lines to allow the reader to easily understand the formulas – and to adjust their values to local conditions if necessary. Although the amount of data on which some of these curves are based is sometimes too insignificant to be statisticaly relevant, the formulas have been applied successfully and adjusted on the basis of practical experience.

The formulas are simple. Besides basic arithmetical operations, they use only one logical function, namely the "IF"-function. For example:

> If temperature is less than 20°C; then hydraulic retention time is 20 days; if not, then it is 15 days in case the temperature is less than 25°C; otherwise (this means, if temperature is over 25°C) the HRT is 10 days.

Assuming the temperature is stated in cell F5 of the spreadsheet, the formula for retention time will be written as: =IF(F5<20;20;IF(F5<25;15;10)).

The formulas have been kept simple, so that the user can make modifications, according to experience or superior knowledge. For example, if it has been found that, for a certain substrate, the HRT should be 25 days below 20°C, 23 days up to 25°C and 20 days above 25°C and, that for safety reasons, 10% longer retention time is added, then the formula should read:

= 110%* IF (F5 < 20; 25; IF (F5 < 25; 23; 20)).

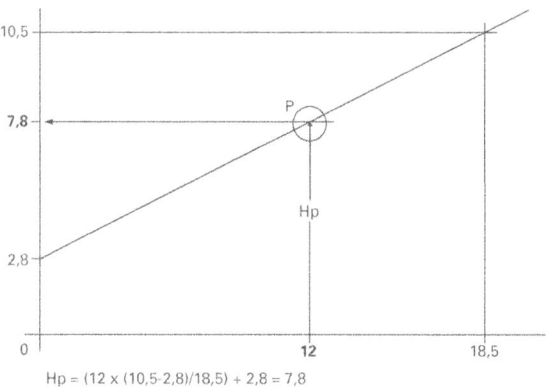

Picture 10_1:
The graphical expression of the "rule of three" for solving proportions: if we know three numbers a, b, and c, and want to find d such that a/b = c/d, then d = cb/a.

Hp = (12 x (10,5-2,8)/18,5) + 2,8 = 7,8

Values between defined days or amounts may be calculated by using the famous "rule of three", of which there are plenty of examples in the following tables. The slope of a straight line is expressed in its tangent; the height of a certain point is found by multiplying the length with the ratio of the slope, i.e. total height divided by total length (Picture 10_1).

In case the reader is not familiar with working in EXCEL, it is better not to modify formulas but to manipulate the results by entering "modified" data. For example, if the values of spreadsheet results are generally lower or higher than the experienced in the field, dimensioning can be adjusted by entering lower or higher temperature values, or shorter or longer retention times. One could also multiply wastewater volumes or COD concentrations by a safety factor before starting the calculation. To avoid mishaps, all the spreadsheet cells should be locked, except the ones written in bold figures.

When the user prepares his or her own tables and copies the formulas below, the columns (A; B; C; D...) and the rows (1; 2; 3; 4; 5...) before the "equals" symbol of each formula, define which cell the formula should be written in. The names of cells and rows are shown on the entry mask of the monitor. In transferring formulas to the spreadsheet, the cell name before the "equals" symbol should not be copied; for example E6=D5/E5 should be written in cell E6 as =D5/E5.

The italic figures are either guiding figures to show usual values, or they indicate limits to be observed. The bold figures are those which have to be filled in by hand; the other figures are calculated. Columns which are labelled "given" contain data which reflects a given reality, for example, wastewater-flow volume or wastewater strength. Columns which are labelled "chosen" contain data which may be modified to optimise the design, for example, hydraulic retention time or desludging intervals. All other cells contain formulas and should be locked, in order to avoid deleting by accident formulas. Cells which are labelled "check" or "require" should be used to confirm whether the chosen and given values are realistic.

10.2 Technical spreadsheets – application

10.2.1 Assumed COD/BOD ratio

The COD/BOD ratio widens during biological treatment because the BOD reflects only that part of the oxygen demand which is reduced by biological treatment, while the COD represents total oxygen demand. The removed BOD, therefore, has a greater percentage-wise influence on the change of the BOD than on the COD. The COD/BOD ratio widens faster while biological degradation is incomplete, and slower when treatment efficiency reaches almost 100%.

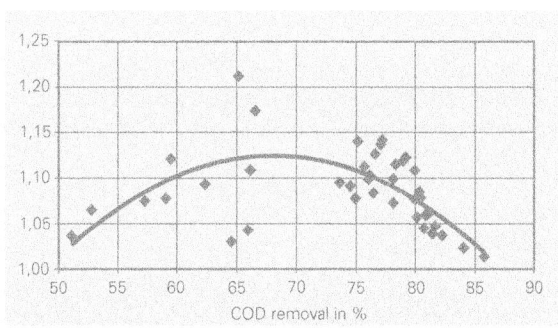

Figure 10_2:
COD removal in relation to temperature in anaerobic reactors.
Change of COD/BOD ratio during anaerobic treatment. The samples have been taken by SIITRAT from anaerobic filters, most of them serving schools in the suburbs of Delhi, India

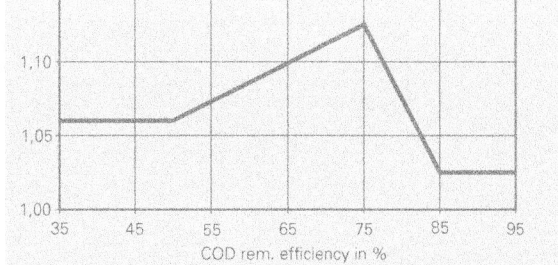

Figure 10_3:
COD removal relative to wastewater strength in anaerobic filters.
Simplified curve of Picture 10_2, which is used in the spreadsheet formulas

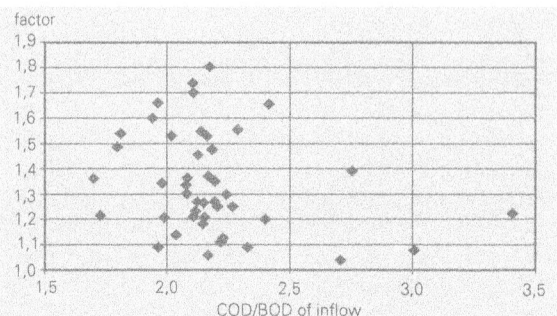

Picture 10_4:
Illustration to spreadsheet for calculation of anaerobic filter dimensions. Changes of COD/BOD ratio during anaerobic treatment of domestic wastewater. The samples were taken by SIITRAT. The sample points of high COD/BOD (to the right of the graph) stem from post-treatment and are not comparable to the majority of samples

10.2.2 Domestic wastewater quantity and quality

The following spreadsheet (see Table 24) helps to define domestic wastewater in terms of the number of people and the wastewater they discharge. BOD and water-consumption figures vary widely from place to place and, therefore, should be obtained for each site.

Formulas of spreadsheet "wastewater per capita":

$E5 = A5 \times C5 / 1000$

$F5 = A5 \times B5 / E5$

$G5 = D5 \times F5$

	A	B	C	D	E	F	G
1	Wastewater production per capita							
2	user	BOD_5 per user	water consump. per user	COD/BOD_5 ratio	daily flow of wastewater	BOD_5 concentr.	COD concentr.
3	given	given	given	given	calculated	calculated	approx
4	number	g/day	litres/day	mg/l	m³/day	mg/l	mg/l
5	80	55	165	1.90	13.20	333	633
6	range =>	40 - 65	50 - 300				

Table 24:
Spreadsheet for calculation of quantity and quality domestic-wastewater production

10.2.3 Septic tank

The size of septic tanks is standardised in most countries. In the case of DEWATS, however, the wastewater may not be the standard domestic wastewater. The spreadsheet (see Table 25) assists in the design of septic tanks. Flow volume, number of peak hours of flow and pollution load are the basic entries. "Chosen" parameters include the desludging interval and the HRT; the former is decisive for the digester volume for sludge storage, while the latter decides the volume of the liquid.

As sludge compacts with time, the formulas in the spreadsheets are based on the graph shown in Picture 10_5.

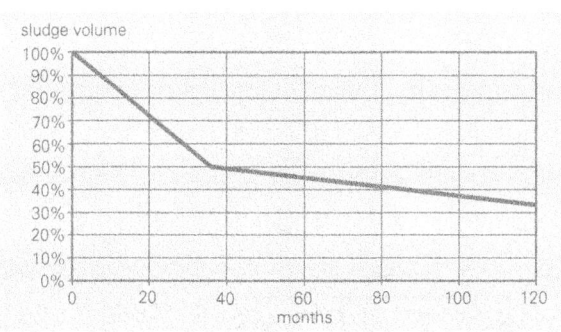

Picture 10_5:
Reduction of sludge volume during storage

COD-removal rates in settlers and septic tanks depend on the amount of settleable solids, their COD content and the intensity of inoculation with fresh inflow. The contact between fresh incoming substrate and active sludge in Imhoff tanks is nearly zero, while in sedimentation ponds with a deep inlet, it is intensive. This fact has been taken into consideration by dividing the parameter "settleable SS per COD" by an experience factor of between 0.50 and 0.60. The general tendency is shown in the graph Picture 10_6.

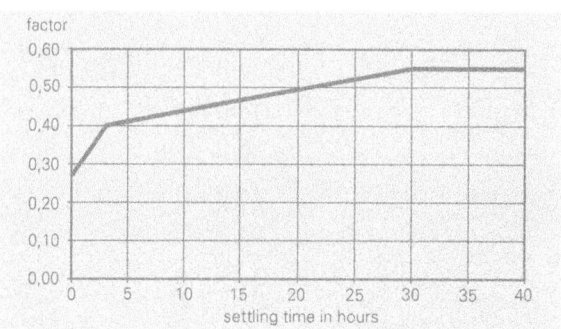

Picture 10_6: COD removal in settlers

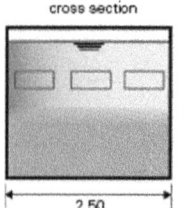
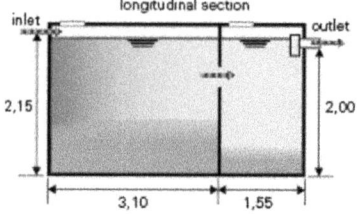

Picture 10_7: Illustration to spreadsheet for calculation of sep tank dimensions

10 Designing DEWATS

Formulas of spreadsheet "septic tank"

A	B	C	D	E	F	G	H	I	J
General spreadsheet for septic tank, input and treatment data									
daily waste-water flow	time of most waste water flow	max. flow at peak hours	COD inflow	BOD_5 inflow	HRT inside tank	settleable SS/COD ratio	COD removal rate	COD outflow	BOD_5 outflow
given	given	calcul.	given	given	chosen	given	calcul.	calcul.	calcul.
m³/day	h	m³/h	mg/l	mg/l	h	mg/l	%	mg/l	mg/l
13.0	12	1.08	633	333	18	0.42	35%	411	209
		COD/BOD_5 ->	1.90		12 - 24 h	0.35 - 0.45 domestic		BOD rem. ->	1.06
dimensions of septic tank									
desludging interval	inner width of septic tank	min. water depth at outlet point	inner length of first chamber		length of second chamber		volume incl. sludge	actual volume of septic tank	biogas 70% CH_4 50% dissolved
chosen	chosen	chosen	requir.	chosen	requir.	chosen	requir.	check	calcul.
months	m	m	m	m	m	m	m³	m³	m³/d
12	2.50	2.00	3.13	3.10	1.56	1.55	23.46	23.25	0.72
						sludge l/g BOD rem.	0.0042		

Table 25: Spreadsheet for calculating septic tank dimensions

C5 = A5/B5

H5 = G5 / 0.6 × IF (F5 < 1; F5 x 0.3; IF (F5 < 3; (F51) x 0.1/2 + 0.3; IF (F5 < 30; (F5 - 3) x 0.15/27 + 0.4; 0.55)))

The formula relates to Picture 10_6. The number 0.6 is a correction factor based on practical experience.

I5 = (1 - H5) x D5

J5 = (1 - H5 x J6) x E5

E6 = D5 / E5

J6 = IF (H5 < 0.5; 1.06; IF (H5 < 0.75; (H5 - 0.5) x 0.065 / 0.25 + 1.06; IF (H5 < 0.85;1.125 - (H5 - 0.75) x 0.1 /0.1; 1.025)))

The formula relates to Picture 10_3.

D11 = 2/3 x H11 / B11 / C11

F11 = D11 / 2

H11 = IF (H12 x (E5 - J5) / 1000 x A11 x 30 x A5 + C5 x F5 < 2 x A5 x F5 / 24; 2 x A5 x F5 / 24; H12 x (E5 - J5) / 1000 x A11 x 30 x A5 + C5 x F5) + 0.2 x B11 x E11

The formula takes into account that sludge volume is less than half the total volume.

I11 = (E11 + G11) x C11 x B11

J11 = (D5 - I5) x A5 x 0.35 / 1000 / 0.7 x 0.5

350 l methane are produced from each kg COD removed.

H12 = 0.005 x IF (A11 < 36; 1 - A11 x 0.014; IF (A11 < 120; 0.5 - (A11 - 36) x 0.002; 1/3))

The formula relates to Picture 10_5.

10.2.4 Fully mixed digester

Within a fully mixed digester, or biogas plant, as it is commonly known in rural households in India, cattle dung is thoroughly mixed with water. Even as effluent, the substrate is very viscous; very little sludge settles and, as a result, no sludge must be removed for many years. The same type of rural biogas plant in China receives a substrate which is a mixture of human excreta, pig dung and water – however, less homogeneous by far than in India. Other wastewater, for example from slaughter-houses, may have different properties again. It is difficult, therefore, to calculate dimensions for the many different kinds of "strong" wastewater, for which biogas treatment might be suitable. The following spreadsheet should be used with certain reservations and – formulas may need to be adapted to local conditions.

The spreadsheet does reveal, however, the influencing factors.
The formulas are based on the following assumptions:
- solids which settle within one day of benchmark testing represent 95% of all settleable solids
- there is a mixing effect inside the digester because of the relatively high gas production, which prevents sludge from settling. Any additional sludge will only make up for the loss in volume by compression. Thus, the accumulating sludge volume is equal to the amount calculated from the one day of benchmark testing
- all settleable and non-settleable solids will digest within hydraulic retention times typical for sludge reactors
- 95% of their BOD is removed after 25 days and 30°C; this is equivalent to 400l of biogas produced from 1kg of organic dry matter

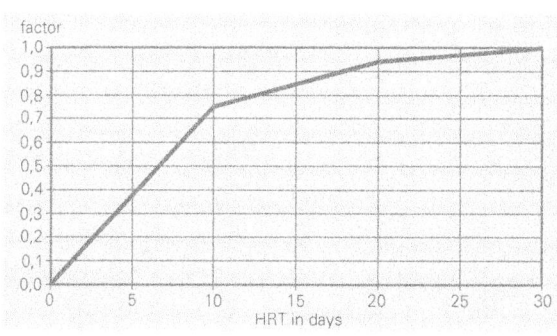

Figure 10_8:
Gas production of fixed-dome biogas plants in relation to HRT

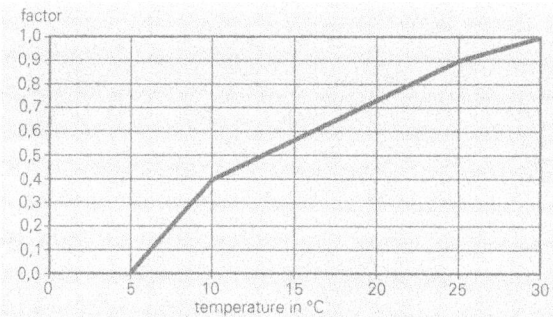

Figure 10_9:
Gas production of fixed-dome biogas plants in relation to temperature

Formulas of spreadsheet "fully mixed digester"

D5 = B5 × C5

I5 = IF (F5 < 10; F5 × 0.75 / 10; IF (F5 < 20; (F5 - 10) × 0.19 / 10 + 0.75; (F5 - 20) × 0.06 / 10 + 0.94))

The formula relates to Picture 10_8.

J5 = IF (G5 < 5; 0; IF (G5 < 10; (G5 - 5) × 0.4 / 5;

IF (G5 < 25; (G5 - 10) × 0.5 / 15 + 0.4; (G5 - 25) × 0.1 / 5 + 0.9)))

The formula relates to Picture 10_9.

K5 = H5 × I5 × J5 × A5 × D5

B11 = 1.1 × ((1000 × K5 × L5 / A11 / 0.35) / (0.95 × I5 × J5)) × (1 - 0.95 × I5 × J5) / A5

The formula determines the influent COD and calculates the COD removal by assuming a production of 350 l methane per kg COD removed; the additional 10% represent the inorganic COD, which is not removed.

D11 = 30 × C11 × A5 × E5 / 1000

E11 = F5 × A5

F11 = D11 + E11

H11 = K5 × G11

L11 = 2 × SQRT ((H11 / J11 - (K11 / 2) × (K11 / 2) × PI()) / PI())

The mathematical expression is:

$$2 \times \sqrt{\frac{\left(\frac{H11}{J11} - \left(\frac{K11}{2}\right)^2 \times \pi\right)}{\pi}}$$

D17 = A17 - B17 / 2

E17 = H11 / (D17 × D17 × PI())

The mathematical expression is:

$H11 / (D17^2 \times \pi)$

$F17 = (F11 - POWER(A7 - B17 - C17; 2) \times PI() \times E17) / (A17 \times A17 \times PI()) + E17$

The mathematical expression is:

$$F11 - \frac{(A17 - B17 - C17)^2 \times \pi \times E17}{A17 \times \pi} + E17$$

$G17 = E17 + 0.15$

$H17 = F17 + 0.15$

$I17 = 3.14 \times I11 \times I11 \times (K17 - I11/3)$

$J17 = 0.02 + POWER((F11 + H11/2 + I17) / 4.19; 1/3)$

The theoretical digester volume is taken as the volume below the zero line plus half the gas storage; 0.02m are added for plaster.

The mathematical expression is:

$0.02 + \sqrt[3]{\frac{(F11 + H11/2 + I17)}{4.19}}$, 4.19 is $4/3\pi$

$L17 = 4.19 \times (K17 - 0.02) \times (K17 - 0.02) \times (K17 - 0.02) - I17 - H11/2$

$B23 = PI() \times (I11 + A23) \times (I11 + A23) \times (K17 - (I11 + A23)/3)$

The volume above the lowest slurry level is found by trial and error; π is expressed as PI().

$C23 = I17 + H11$

$D23 = A23 + J11$

$E23 = 3.14 \times I11 \times I11 \times (G23 - I11/3)$

$F23 = 0.02 + POWER((F11 + H11/2 + E23) / 2.09; 1/3)$

The mathematical expression is:

$$0.02 + \sqrt[3]{\frac{(F11 + H11/2 + E23)}{2.09}} \; ; \; 2.09 \text{ is } 2/3\pi$$

$H23 = 2.09 \times (G23 - 0.02) \times (G23 - 0.02) \times (G23 - 0.02) - E23 - H11 / 2$

$J23 = PI() \times (I23 + I11) \times (I23 + I11) \times (G23 - (I23 + I11) / 3)$

The volume above the lowest slurry level is found by trial and error; π is expressed as PI().

$K23 = E23 + H11$

$L23 = I23 + J11$

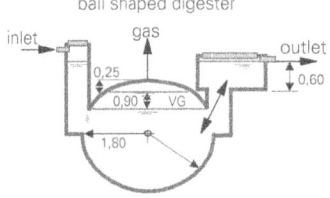

ball shaped digester

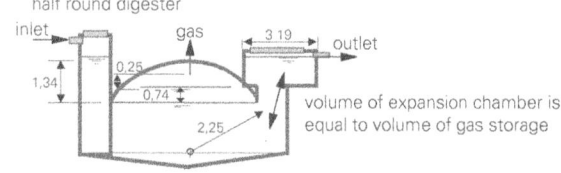

half round digester

volume of expansion chamber is equal to volume of gas storage

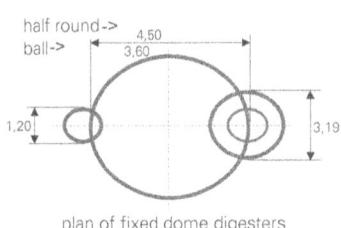

plan of fixed dome digesters

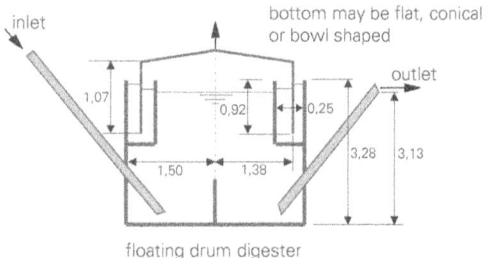

bottom may be flat, conical or bowl shaped

floating drum digester

Picture 10_10: Illustration to spreadsheet for calculation of fully mixed digester dimensions

Designing DEWATS

	A	B	C	D	E	F	G	H	I	J	K	L
1				General spreadsheet for biogas plants, input and gas-production data								
2	daily flow	TS (DM) content	org. DM/ total DM	org. DM content	solids settleable within one day	HRT	lowest digester tempera-ture	ideal biogas product at 30°C	gas production factors		total gas product	methane content
3	given	given	assumed	calcul.	tested	chosen	given	given	calcul. acc. to graphs		calcul.	assumed
4	m³/d	%	ratio	%	ml/l	d	°C	l/kg org DM	f-HRT	f-temp	m³/d	ratio
5	0.60	6.0	67%	4.0	20	25	25	400	0.97	0.90	8.42	70%
6								200 - 450				
7			values for all digester shapes						for all fixed-dome plants			
8	non-dissolv. methane prod.	approx. effluent COD	de-sludging interval	sludge volume	liquid volume	total digester volume	gas storage capacity	gas holder volume VG	free distance above slurry zero line	outlet above zero	dia-meter of left shaft	dia-meter of expans. chamber
9	assumed	calcul.	chosen	calcul.	calcul.	calcul.	given	calcul.	chosen	chosen	chosen	calcul.
10	ratio	mg/l	months	m³	m³	m³	ratio	m³	m	m	m	m
11	80%	7,943	12	4.32	16.0	19.3	65%	5.5	0.25	0.60	1.20	3.19
12										minimum 0.60 m		
13			cylindrical floating-drum plant						ball-shaped digester			
14	radius of digester	width of water ring	wall thickness of water ring	radius of gas holder	theor. height of gas holder	theor. depths of digester	actual heigh-of gas holder	actual depth of digester	volume of empty space above zero line	radius ball shape	actual digester radius (ball)	actual net volu-me of digester
15	chosen	chosen	chosen	calcul.	calcul.	calcul.	calcul.	calcul.	calcul.	requir.	chosen	check
16	m	m	m	m	m	m	m	m	m³	m	m	m³
17	1.50	0.25	0.12	1.38	0.92	3.13	1.07	3.28	0.34	1.77	1.80	20.56
18												
19			ball shaped digester					half-ball shaped digester				
20	lowest slurry level below zero line (fill in trial until "calcul." match "target")		gas pressure ball shaped	volume of empty space above zero line	radius half round shape	actual digester radius (half round)	actual net volume of diges-ter	lowest slurry level below zero line (fill in trial until "calcul." match "target")			gas pressure half-ball	
21	trial!!	calcul.	target	calcul.	calcul.	requir.	chosen	check	trial!!	calcul.	target	calcul.
22	m	m³	m³	m w.c.	m³	m	m	m³	m	m³	m³	m w.c.
23	0.90	5.89	5.81	1.60	0.43	2.23	2.25	20.01	0.74	5.91	5.90	1.34
24				1.50 max.								1.50 max.

Table 26:
Spreadsheet for calculating fully mixed digester dimensions

10.2.5 Imhoff tank

The general treatment properties in the Imhoff tank are comparable to those in any other settler. Since wastewater does not come into direct contact with active sludge, BOD removal from the liquid is almost zero; however, as sedimentation is greater than in other settlers, the COD or BOD removal within these units is comparable. This fact is reflected in the factor 0.50 of cell H5.

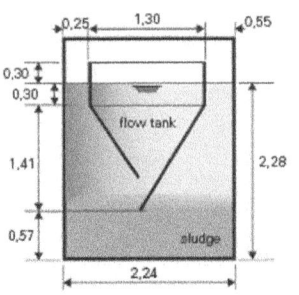
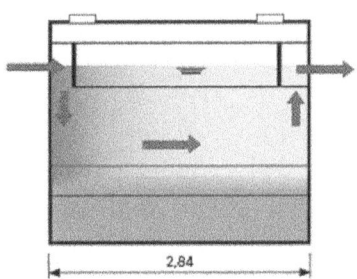

Picture 10_11: Illustration to spreadsheet for calculation of Imhoff Tank dimensions

Designing DEWATS

Flow volume, number of peak hours of flow and pollution load are the basic entries for calculation. "Chosen" parameters are the same as those for the septic tank – HRT and desludging intervals.

Formulas of spreadsheet "Imhoff tank"

	A	B	C	D	E	F	G	H	I	J
1	General spreadsheet for Imhoff tank, input and treatment data									
2	daily waste-water flow	time of most waste-water flow	max. flow at peak hours	COD inflow	BOD_5 inflow	HRT inside flow tank	settleable SS/COD ratio	COD removal rate	COD outflow	BOD_5 outflow
3	given	given	calcul.	given	given	chosen	given	calcul.	calcul.	calcul.
4	m³/day	h	m³/h	mg/l	mg/l	h	mg/l	%	mg/l	mg/l
5	25.0	12	2.08	633	333	1.50	0.42	27%	460	237
6				COD/BOD_5 ->	1.90		domestic 0.35 - 0.45		BODrem. ->	1.08
7	dimensions of Imhoff tank									
8	desludging interval	flow tank volume	sludge volume	inner width of flow tank	space beside flow tank	total inner width of Imhoff tank	inner length of Imhoff tank	sludge height	total depth at outlet	biogas 70% CH_4 50% dissolved
9	chosen	calcul.	calcul.	chosen	chosen	calcul.	calcul.	calcul.	calcul.	calcul.
10	months	m³	m³	m	m	m	m	m	m	m³/d
11	12	3.13	3.61	1.30	0.55	2.24	2.82	0.57	2.28	1.08
12		sludge l/g BODrem	0.0042							

Table 27:
Spreadsheet for calculating Imhoff tank dimensions

$C5 = A5/B5$

$H5 = G5 / 0.5 \times IF (F5 < 1; F5 \times 0.3; IF (F5 < 3; (F5 - 1) \times 0.1 / 2 + 0.3 ;$
$IF (F5 < 30; (F5 - 3) \times 0.15 / 27 + 0.4; 0.55)))$

The formula relates to 10_6. The number 0.5 is a correction factor based on practical experience.

$I5 = (1 - H5) \times D5$

$J5 = (1 - H5 \times J6) \times E5$

$E6 = D5 / E5$

$J6 = IF (H5 < 0.5; 1.06; IF (H5 < 0.75; (H5 - 0.5) \times 0.065 / 0.25 + 1.06;$
$IF (H5 < 0.85; 1.125 - (H5 - 0.75) \times 0.1 / 0.1; 1.025)))$

The formula relates to Picture 10_3.

$B11 = C5 \times F5$

$C11 = A5 \times 30 \times A11 \times C12 \times (E5 - J5) / 1000$

$F11 = D11 + E11 + 0.25 + 2 \times 0.07$

All formulas for dimensions relate to the geometry of the Imhoff tank, as shown in Picture 10_11.

$G11 = B11 / (0.3 \times D11 + (D11 \times D11 \times 0.85 / 2))$

$H11 = C11 / F11 / G11$

$I11 = H11 + 0.85 \times D11 + 0.3 + 0.3$

$J11 = (D5 - I5) \times A5 \times 0.35 / 1000 / 0.7 \times 0.5$

350 l methane are produced from each kg COD removed.

$C12 = 0.005 \times IF (A11 < 36;1 - A11 \times 0.014; IF (A11 < 120; 0.5 - (A11 - 36) \times 0.002;1/3))$

The formula relates to Picture 10_5.

10.2.6 Anaerobic baffled reactor

Volume of flow, number of peak hours of flow and pollution load are the basic entries for calculation. "Chosen" parameters for designing a baffled reactor are the HRT, desludging intervals and the up-flow velocity (cell I17). Due to the interrelation between these factors, the HRT cannot be reduced by changing the dimensions of the up-flow chambers because the up-flow velocity will thereby be increased. To achieve the desired effluent quality, it is better to add another chamber than to enlarge their volumes because treatment efficiency increases with the number of chambers (see formula of cell J17). However, practical experience has shown that treatment efficiency does not increase with more than six chambers. Calculation is based on the curve (Picture 10_12) showing BOD removal for a BOD of 900mg/l at 25°C. Factors are applied to adapt the calculation to waste-water strength (Picture 10_14) and temperature (Picture 10_17). An additional curve is used to prevent organic overloading (Picture 10_13).

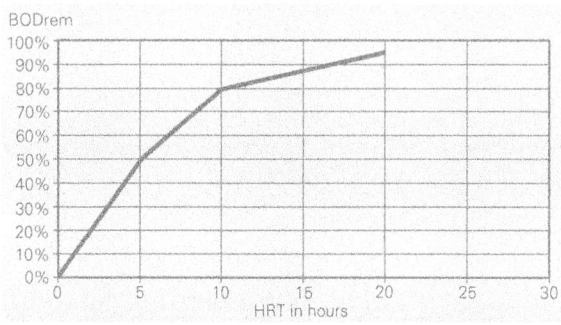

Picture 10_12: BOD removal in relation to HRT in baffled reactors

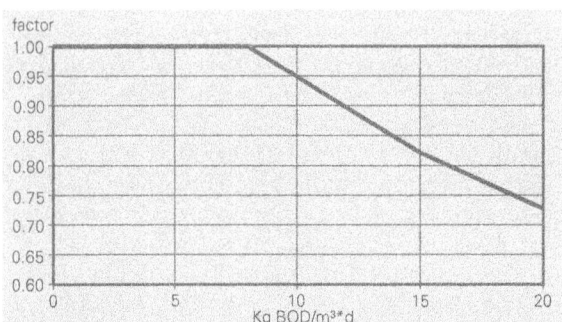

Picture 10_13:
BOD removal affected by organic overloading baffled reactors

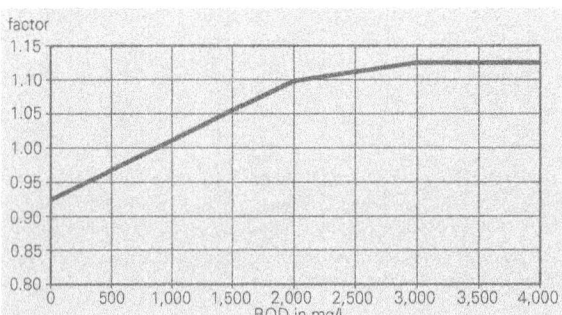

Picture 10_14:
BOD removal in baffled reactors in relation to waste water strength

0 Designing DEWATS

	A	B	C	D	E	F	G	H	I	J	K
1	General spreadsheet for ABR with integrated settler										
2	daily waste-water flow	time of most waste-water flow	max. peak flow per hour	COD inflow	BOD_5 inflow	COD/BOD ratio	settleable SS / COD ratio	lowest digester temperature	de-sludging interval	HRT in settler (no settler HRT=0)	COD removal rate in settler
3	avg.	given	max.	given	given	calcul.	given	given	chosen	chosen	calcul.
4	m^3/day	h	m^3/h	mg/l	mg/l	ratio	mg/l	°C	months	h	%
5	25	12	2.08	633	333	1.90	0.42	25	18	1.50	23%
6			COD/BOD_5 ->				0.35-0.45			1.5 h	
7	treatment data										
8	BOD_5 removal rate in settler	inflow into baffled reactor		COD/BOD_5 ratio after settler	factors to calculate COD removal rate of anaerobic filter			COD rem. 25°, COD 1500	theor. rem. rate acc. to factors	COD rem. rate baffle only	COD out
9	calcul.	COD	BOD_5	calcul.	calculated according to graphs			calcul.	calcul.	calcul.	
10	%	mg/l	mg/l	mg/l	f-overload	f-strenght	f-temp	f-HRT %	%	%	mg/l
11	24%	489	253	1.94	1.00	0.91	1.00	87	79%	81%	94
12	1.06	<- COD/ BOD removal factor							COD/BOD removal factor ->		1.025
13	dimensions of settler								ABR		
14	total COD removal rate	total BOD_5 removal rate	BOD_5 out	inner masonry measurements chosen acc. to required volume		sludge accum. rate	length of settler	length of settler	max. upflow velocity	number of upflow chambers	depth at outlet
15	calcul.	calcul.	calcul.	width	depth	calcul.	calcul.	chosen	chosen	chosen	chosen
16	%	%	mg/l	m	m	l/g COD	m	m	m/h	No.	m
17	85%	87%	42	2.00	1.50	0.0037	2.39	2.40	1.8	5	1.50
18									1.4-2.0 m/h		
19	dimensions of ABR							status and gp			
20	length of chambers should not exceed half depth	area of single upflow chamber	width of chambers		actual upflow velocity	width of downflow shaft	actual volume of baffled reactor	actual total HRT	org. load (BOD_5)	biogas (ass: CH_4 70%; 50% dissolved)	
21	calcul.	chosen	calcul.	calcul.	chosen	calcul.	chosen	calcul.	calcul.	calcul.	calcul.
22	m	m	m^2	m	m	m/h	m	m^3	h	kg/m^3·d	m^3/d
23	0.75	0.75	1.16	1.54	2.00	1.39	0.25	15.00	14	1.63	3.37
24									HRT reduced by 5% for sludge		

TIP: If removal rate is insufficient, increase number of upflow chambers to keep upflow velocity low.

Table 28
Spreadsheet for the calculation of anaerobic baffled reactor dimensions

Formulas of spreadsheet "ABR"

C5 = A5 / B5

F5 = D5 / E5

K5 = G5 / 0.6 x IF (J5 < 1; J5 x 0.3; IF (J5 < 3; (J5 - 1) x 0.1/2 + 0.3;
IF (J5 < 30; (J5 - 3) x 0,15 / 27 + 0.4;0.55)))

The formula relates to Picture 10_6. The number 0.6 is a correction factor based on practical experience.

A11 = K5 x A12

B11 = D5 x (1 - K5)

C11 = E5 x (1 - A11)

D11 = B11 / C11

E11 = IF (J23 < 8;1; IF (J23 < 15;1 - (J23 - 8) x 0.18 / 7; 0,82 - (J23 - 15) x 0.9 / 5))

The formula relates to Picture 10_13.

F11 = IF (B11 < 2000; B11 x 0.17 / 2000 + 0.87;
IF (B11 < 3000; (B11 - 2000) x 0.02 / 1000 + 1.04; 1.06))

The formula relates to Picture 10_14.

G11 = IF (H5 < 20; (H5 - 10) x 0.39 / 20 + 0.47; IF (H5 <25; (H5 - 20) x 0.14 /5 + 0.86;
IF(H5<30;(H5-25)x0.08/5+1;1.1)))

The formula relates to Picture 10_17.

H11 = IF(I23 < 5; I23 x 0.51 / 5; IF (I23 < 10; (I23 - 5) x 0.31 /5 + 0.51;
IF (I23 < 20; (I23 - 10) x 0.13 / 10 + 0.82; 0.95)))

I11 = E11 x F11 x G11 x H11

The formula relates to Picture 10_12.

J11 = IF (J17 < 7; E11 x F11 x G11 x H11 x (J17 x 0.04 + 0.82); E11 x F11 x G11 x H11 x 0.98)

The formula considers improved treatment by increasing the number of chambers and limiting the treatment efficiency to 98%.

K11 = (1 - J11) x B11

A12 = IF (K5 < 0.5; 1.06; IF (K5 < 0.75; (K5 - 0.5) x 0.065 / 0.25 + 1.06;
IF (K5 < 0.85; 1.125 - (K5 - 0.75) x 0.1 / 0.1; 1.025)))

The formula relates to Picture 10_3.

K12 = IF (A17 < 0.5; 1.06; IF (A17 < 0.75; (A17 - 0.5) x 0.065 / 0.25 + 1.06;
IF (A17 < 0.85; 1.125 - (A17 - 0.75) x 0.1 /0.1; 1.025)))

The formula relates to Picture 10_3.

A17 = 1 - K11/D5

B17 = A17 x K12

C17 = (1 - B17) x E5

F17 = 0.005 x IF(I5<36;1-I5x0.014;IF(I5<120;0.5-(I5-36)x0.002;1/3))

The formula relates to Picture 10_5.

G17 = IF (A11 > 0; IF (F17 x (E5 - C11) / 1000 x 30 x I5 x A5 + J5 x C5 < 2 x J5 x C5;
2 x J5 x C5; F17 x (E5 - C11) / 1000 x 30 x I5 x A5 + J5 x C5); 0) / D17 / E17

The formula considers that sludge volume is less than half of the total volume; a settler may be omitted.

A23 = K17 x 0.5

C23 = C5 / I17

D23 = C23 / B23

F23 = C5 / B23 / E23

H23 = (G23 + B23) x J17 x K17 x E23

I23 = H23 / (A5 / 24) / 105%

J23 = B11 x C5 x 24 / H23 / 1000

K23 = (D5 - K11) x A5 x 0.35 / 1000 / 0.7 x 0.5

350l methane are produced from each kg COD removed.

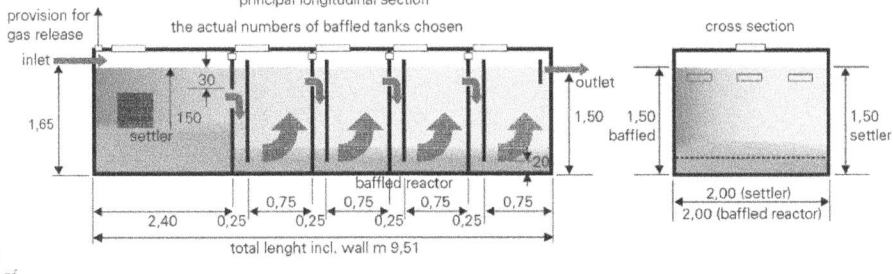

Figure 10_15
Illustration
spreadsheet
calculation of
ABR dimensions

10.2.7 Anaerobic filter

Volume of flow and pollution load are the basic entries for calculation. The "chosen" parameters for the anaerobic filter are the hydraulic retention time and desludging intervals. The calculation of performance is based on a curve, which describes the relation between hydraulic retention time and percentage of COD removal. The curve (Picture 10_16) is based on a COD of 1500mg/l at 25°C. The values are then multiplied by factors reflecting temperature (Picture 10_17), wastewater strength (Picture 10_18) and specific-filter surface (Picture 10_19).

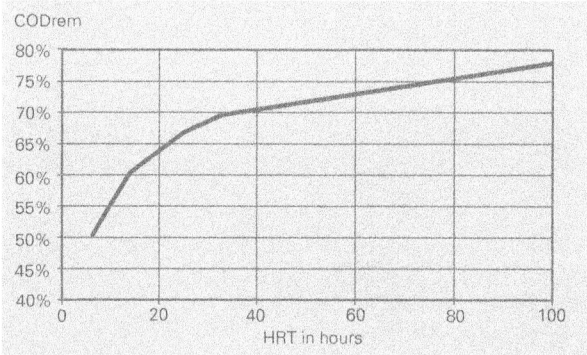

Picture 10_16: COD removal in relation to HRT anaerobic filters

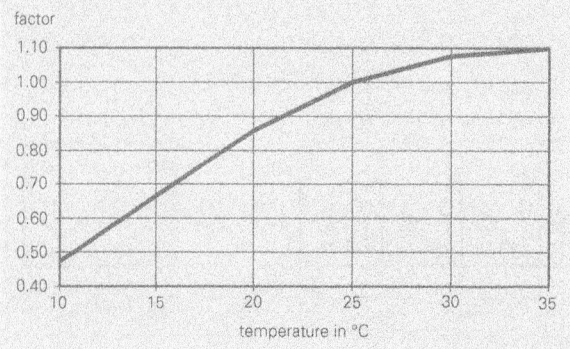

Picture 10_17: COD removal in relation to temperature in anaerobic reactors

The void space of the filter medium influences the digester volume required to provide sufficient hydraulic retention time. Gravel has approximately 35% void space, while specially manufactured plastic pieces may have over 90%. When filter height is increased together with total water depth, the impact of increased depth on HRT is less with gravel than with plastic pieces. While filter height remains the same, the distance from filter bottom to digester floor must be increased.

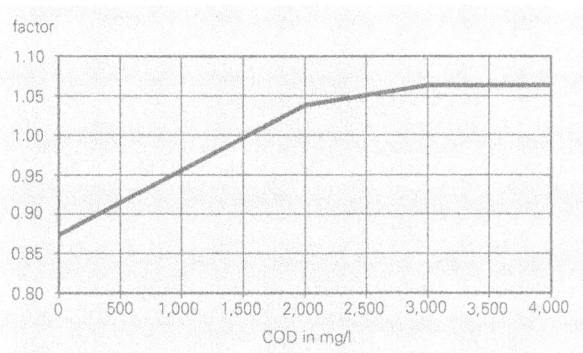

Picture 10_18: COD removal relative to wastewater strength in anaerobic filters

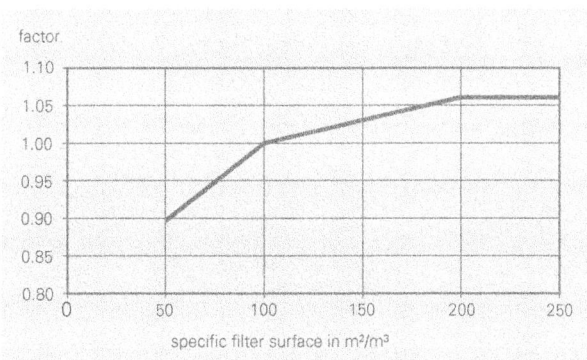

Picture 10_19: Spreadsheet for calculating anaerobic-filter dimensions

	A	B	C	D	E	F	G	H	I	J	K	L
1	General spreadsheet for anaerobic filter (AF) with integrated septic tank (ST)											
2	daily waste water flow	time of most waste water flow	max. peak flow per hour	COD inflow	BOD_5 inflow	$SS_{settl.}/$ COD ratio	lowest digester temperature	HRT in septic tank	de-sludging interval	COD-removal septic tank	BOD_5 removal septic tank	BOD/COD-removal factor
3	given	given	calcul.	given	given	given	given	chosen	chosen	calcul.	calcul.	calcul.
4	m^3/day	h	m^3/h	mg/l	mg/l	mg/l	°C	h	month	%	%	ratio
5	25.0	12	2.08	633	333	0.42	25	2	36	25%	26%	1.06
6			COD/BOD_5 ->	1.90	0.35 - 0.45 (domestic)			2h			BOD_{min} -> 1.06	
7	treatment data											
8	COD inflow in AF	BOD_5 inflow into AF	specific surface of filter medium	voids in filter mass	HRT inside AF reactor	factors to calculate COD-removal rate of anaerobic filter				COD-removal rate (AF only)	COD outflow of AF	COD-removal rate of total system
9	calcul.	calcul.	given	given	chosen	calculated according to graphs				calcul.	calcul.	calcul.
10	mg/l	mg/l	m^2/m^3	%	h	f-temp	f-strength	f-surface	f-HRT	%	mg/l	%
11	478	247	100	35%	30	1.00	0.91	1.00	69%	70	142	78%
12			80 - 120	30 - 45	24 - 48 h							
13	dimensions of septic tank											
14	BOD/COD removal factor	BOD_5 rem. rate of total system	BOD_5 outflow of AF	inner width of septic tank	min. water depth at inlet point	inner length of first chamber		length of second chamber		sludge accum.	volume incl. sludge	actual volume of septic tank
15	calcul.	calcul.	calcul.	chosen	chosen	calcul.	chosen	calcul.	chosen	calcul.	requir.	calcul.
16	ratio	%	mg/l	m	m	m	m	m	m	l/kg BOD	m^3	m^3
17	1.10	85	49	1.75	2.25	1.69	1.70	0.85	0.85	0.00	10.00	10.04
18												sludge l/g BODrem
19	dimension of an aerobic filter							bio gas production			check!	
20	volume of filter tanks	depth of filtertanks	length of each tank	number of filter tanks	width of filter tanks	space below perforated slabs	filter high (top 40cm below water-level)	out of septic tank	out of anaerobic filter	total	org. load on filter volume COD	max. up-flow velocity inside filter voids
21	calcul.	chosen	calcul.	chosen	requir.	chosen	calcul.	assump.	70% CH_4; 50% dissolved		calcul.	calcul.
22	m^3	m	m	No.	m	m	m	m^3/d	m^3/d	m^3/d	kg/m^3*d	m/h
23	31.25	2.25	2.25	3	2.69	0.60	1.20	0.97	2.10	3.07	1.57	0.98
24				max.!!							< 4.5	< 2.0

Table 29:
Spreadsheet for calculating anaerobic-filter dimensions

Formulas of spreadsheet "anaerobic filter"

C5 = A5 / B5

J5 = F5 / 0.6 x IF (H5 < 1;H5 x 0.3; IF (H5 < 3; (H5 - 1) x 0.1 / 2 + 0.3;
IF(H5 < 30; (H5 - 3) x 0.15 / 27 + 0.4; 0.55)))

The formula relates to Picture 10_6. The number 0.6 is a correction factor based on practical experience.

K5 = L5 x J5

L5 = IF (J5 < 0.5; 1.06; IF (J5 < 0.75; (J5 - 0.5) x 0.065 / 0.25 + 1.06;
IF (J5 < 0.85; 1.125 - (J5 - 0.75) x 0.1/0.1; 1.025)))

The formula relates to Picture 10_3.

D6 = D5 / E5

A11 = D5 x (1 - J5)

B11 = E5 x (1 - K5)

F11 = IF (G5 < 20; (G5 - 10) x 0.39 / 20 + 0.47; IF(G5<25; (G5 - 20) x 0.14 / 5 + 0.86;
IF (G5 < 30; (G5 - 25) x 0.08 / 5 + 1;1.1)))

The formula relates to Picture 10_17.

G11 = IF (A11 < 2000; A11 x 0.17 / 2000 + 0.87;
IF (A11 < 3000; (A11 - 2000) x 0.02 / 1000 + 1.04; 1.06))

The formula relates to Picture 10_18.

H11 = IF (C11 < 100; (C11 - 50) x 0.1 / 50 + 0.9; IF (C11 < 200; (C11 - 100) x 0.06 / 100 + 1; 1.06))

The formula relates to Picture 10_19.

I11 = IF (E11 < 12; E11 x 0.16 / 12 + 0.44; IF (E11 < 24; (E11 - 12) x 0.07 /12 + 0.6;
IF (E11 < 33; (E11 - 24) x 0.03 / 9 + 0.67; IF (E11 < 100; (E11 - 33) x 0.09 / 67 + 0.7; 0.78))))

The formula relates to Picture 10_16.

$J11 = IF(F11 \times G11 \times H11 \times I11 \times (1 + (D23 \times 0.04)) < 0.98;$
$F11 \times G11 \times H11 \times I11 \times (1 + (D23 \times 0.04)); 0.98)$

The formula considers improved treatment by increasing the number of chambers and limiting the treatment efficiency to 98%.

$K11 = A11 \times (1 - J11)$

$L11 = (1 - K11 / D5)$

$A17 = IF(L11 < 0.5; 1.06 ; IF(L11 < 0.75; (L11 - 0.5) \times 0.065 / 0.25 + 1.06;$
$IF(L11 < 0.85; 1.125 - (L11 - 0.75) \times 0.1 / 0.1; 1.025)))$

The formula relates to Picture 10_3.

$B17 = L11 \times A17$

$C17 = (1 - B17) \times E5$

$F17 = 2/3 \times K17 / D17 / E17$

$H17 = F17 / 2$

$J17 = 0.005 \times IF(I5 < 36; 1 - I5 \times 0.014; IF(I5 < 120; 0.5 - (I5 - 36) \times 0.002; 1/3))$

The formula relates to Picture 10_5.

$K17 = IF(OR(K5 > 0; J5 > 0); IF(J17 \times (E5 - B11) / 1000 \times I5 \times 30 \times A5 + H5 \times C5 < 2 \times H5 \times C5;$
$2 \times H5 \times C5; J17 \times (E5 - B11) / 1000 \times I5 \times 30 \times A5 + H5 \times C5); 0)$

The formula considers that the sludge volume is less than half of the total volume; a settler may be omitted.

L17 = (G17 + I17) x E17 x D17

A23 = E11 x A5 / 24

C23 = B23

E23 = A23 / D23 / ((B23 x 0.25) + (C23 x (B23 - G23 x (1 - D11))))

G23 = B23 - F23 - 0.4 - 0.05

H23 = (D5 - A11) x A5 x 0.35 / 1000 / 0.7 x0.5

350l methane are produced from each kg COD removed.

I23 = (A11 - K11) x A5 x 0.35 / 1000 / 0.7 x 0.5

350l methane are produced from each kg COD removed.

J23= SUM (H23 : I23)

K23 = A11 x A5 / 1000 / (G23 x E23 x C23 x D11 x D23)

L23 = C5 / (E23 x C23 x D11)

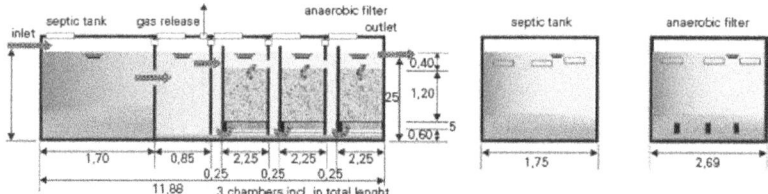

Figure 10_20:
Illustration to spreadsheet for calculating anaerobic-filter

10.2.8 Horizontal gravel filter

Average flow and pollution load are the basic entries for calculation. The "chosen" parameter for the design of gravel filters is the desired effluent quality (BODout, cell E5). The hydraulic retention time and temperature have the greatest influence on treatment performance. The HRT depends on the desired BOD-removal rate (Picture 10_22); the curve is based on 25°C and 35% pore space. The pore space inside the filter defines the "real" HRT, also influenced by the type and number of plants chosen. Further influencing factors are close to 1.0; the information needed to define these factors probably not available at the site anyway.

In practice, the limiting factors are the organic load and the hydraulic load. The limit for hydraulic loading is approximately 100l/m² or 0.1m, although this value can be much higher when using coarse-filter media with guaranteed conductivity. A horizontal filter should not receive more than 10g BOD/(m²xd) because oxygen supply via the surface is limited; this value is only half of the limit for aerobic ponds. This is because a gravel filter works more like a plug flow system; the organic load is much higher in the front section compared to the rear, while oxygen supply is also inferior in the lower part. As a result, the cross-sectional area at the inflow side is influenced by organic loading (cell E12).

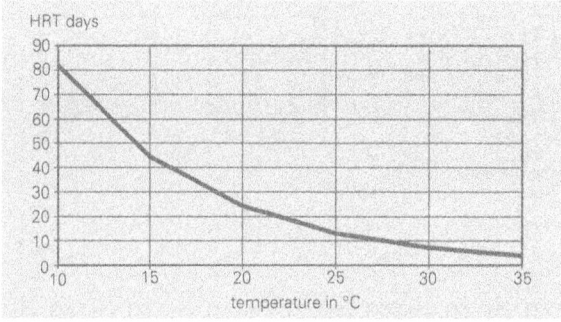

Picture 10_21: HRT relative to temperature in gravel filters, based on 90% BOD-removal

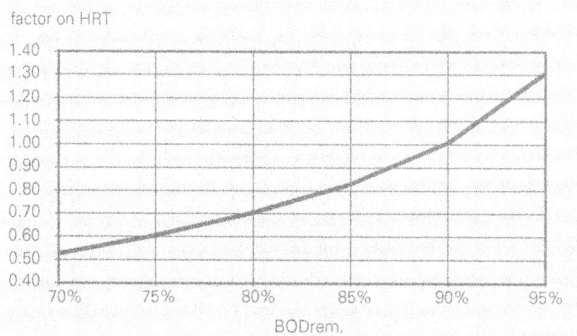

Picture 10_22:
Influence of desired BOD removal rates on HRT of gravel filters, based on 35% pore space at 25°C

Formulas of spreadsheet "gravel filter"

D5 = B5 / C5

F5 = 1 - E5 / C5

G5 = F5 / G6

H5 = B5 x (1-G5)

J5 = IF (F5 < 0.4); (F5 x 0.22) / 0.4); IF (F5 < 0.75; (F5 - 0.4) x 31 / 35 + 0.22;
IF (F5 < 0.8; (F5 - 0.75) x 9.5 / 5 + 0.605; IF (F5 < 0.85; (F5 - 0.8) x 12.5 / 5 + 0.7;
IF (F5 < 0.9; (F5 - 0.85) x 17.5 / 5 + 0.825; (F5 - 0.9) x 30 / 5 + 1))))

The formula refers to Picture 10_22.

K5 = J5 x IF (I5 < 15;82 - (I5 - 10) x 37 / 5; IF (I5 < 20; 45 - (I5 - 15) x 31 / 5;
IF (I5 < 25; 24 - (I5 - 20) x 11 / 5; IF (I5 < 30; 13 - (I5 - 25) x 6 / 5; 7))))

The formula refers to Picture 10_21.

G6 = IF (F5 < 0.5; 1.06; IF (F5 < 0.75; (F5 -0.5)x0.065/0.25+1.06;
IF(F5<0.85;1.125-(F5-0.75)x0.1/0.1;1.025)))

The formula refers to Picture 10_3.

L6 = L5 / 86400

A11 = K5 × 35%

D11 = IF (A5 / L5 / B11 < A5 × C5 / E12; A5 × C5 / E12; A5 / L5 / B11)

The formula compares hydraulic load to maximum organic load in cell E12.

E11 = D11 / C11

F11 = IF (A5 × C5 /L12 > A5 × K5 / C11; A5 × C5 /L12; A5 × K5 / C11)

The formula compares permitted hydraulic load with organic load in cell L12.

G11 = F11 / E11

J11 = H11 × I11

K11 = A5 / J11

L11 = K11 × C5

H12 = E11

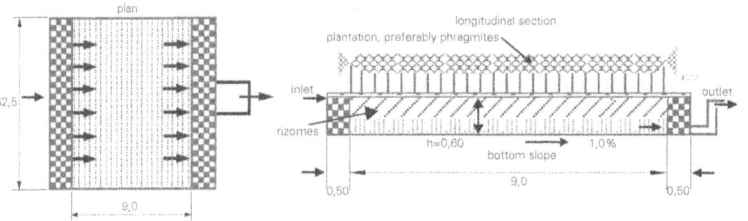

Picture 10_23: Illustration to spreadsheet for calculating dimensions horizontal gravel filter

10 Designing DEWATS

	A	B	C	D	E	F	G	H	I	J	K	L
1	General spreadsheet for planted gravel filter, input and treatment data											
2	average flow	COD in	BOD_5 in	COD/ BOD ratio	outflow BOD_5	BOD_5 removal rate	COD removal	COD out	min. annual Temp.	HRT factor acc. to k20=0.3	HRT	hydraulic conduct. K_s
3	given	given	given	calcul.	wanted	calcul.	calcul.	calcul.	given	calcul. via graph	calcul.	given
4	m³/d	mg/l	mg/l	mg/l	mg/l	%	%	mg/l	°C		days	m/d
5	26	410	215	1.91	30	86%	84	66	25	0.86	11.20	200
6				COD/BOD rem. factor via graph ->			1.025			Ks in m/s ->		2.31E - 0.3
7	dimensions								results			
8	HRT in 35% pore space	bottom slope	depth of filter at inlet	cross section area	width of filter basin	surface area required	length of filter basin	chosen width	lenght chosen	actual surface area chosen	hydr. load on chosen surface	org. load on chosen surface
9	calcul.	chosen	chosen	calcul.	calcul.	calcul.	calcul.	chosen	chosen	check!	calcul.	calcul.
10	days	%	m	m²	m	m²	m	m	m	m²	m/d	g/m² BOD
11	3.92	1.0%	0.60	37.27	62.1	559	9.0	62.5	9.0	563	0.046	9.9
	^ information only	0.3 - 0.6 m		max. BOD_5 150 g/m²			always -> 62.1			max. loads =>	0.100	10

Table 30:
Spreadsheet for calculating
dimensions horizontal gravel filters

10.2.9 Anaerobic pond

Anaerobic ponds should be built for sedimentation purposes only, as highly loaded ponds with very short retention times and heavy scum formation on the surface – or as relatively low-loaded ponds which are almost odourless because of neutral pH values. The spreadsheet may be used for all three categories. The hydraulic retention time is, therefore, the "chosen" parameter. Ponds with long retention times (low organic loading rates) may be divided into several ponds in series. For ponds with short retention times, the front section can be separated to support development of scum. The choice of HRT strongly influences the organic load of the effluent. Ambient temperature is important and an excessively high temperature should not be chosen for want of smaller ponds. It is assumed that temperature has no influence on COD removal at short retention times of less than 30 hours.

Cell G11 should be observed and compared with F11 when the pond is near residential houses.

The biogas potential is also calculated to decide whether a closed anaerobic tank with biogas collection should be built instead.

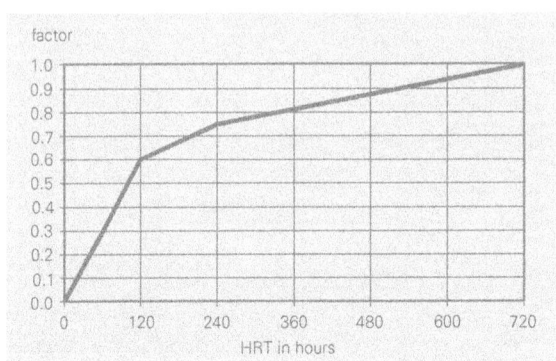

Picture 10_24: Influence of HRT on COD removal non-settled solid anaerobic ponds

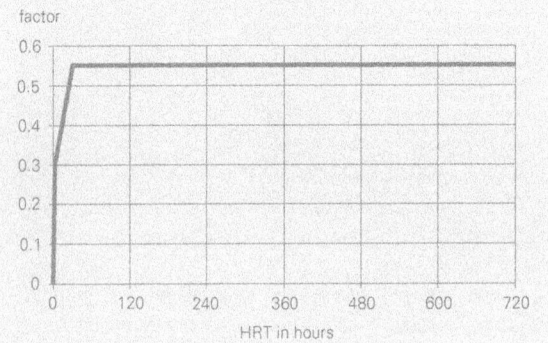

Picture 10_25: Influence of HRT on COD removal of settled solids anaerobic ponds

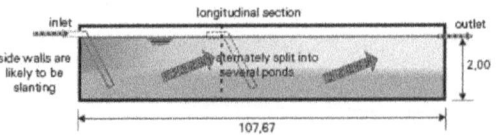 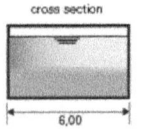

Picture 10_26: Illustration to spreadsheet for calculating dimensions of anaerobic ponds (figures of Table 31)

Formulas of spreadsheet "anaerobic and sedimentation pond"

D5 = B5 / C5

H5 = IF (E5 < 1; F5 / 0.6 x (0.3 x E5); IF (E5 < 3; F5 / 0.6 x (E5 - 1) x 0.1 / 2;
IF (E5 < 30; F5 / 0.6 x ((E5 - 3) x 0.15 / 27 + 0.4);
IF(E5 < 120; E5 x 0.5 x (1 - 0.55 x F5 / 0.6) / 120 + 0.55 x F5 / 0.6;
IF (E5 < 240; (E5 - 120) x 0.25 x (1 - 0.55 x F5 / 0.6) / 120 + 0.5 x (1 - 0.55 x F5 / 0.6) + 0.55 x F5 / 0.6;
IF (E5 < 480; (E5 - 240) x 0.19 x (1 - 0.55 x F5 / 0.6) / 240 + 0.55 x F5 / 0.6 + 0.75 x (10.55 x F5 / 0.6);
(E5 - 480) x 0.06 x (1 - 0.55 x F5 / 0.6) / 240 + 0.55 x F5 / 0.6 + 0.94 x (1 - 0.55 x F5 / 0.6))))))

The formula refers to Picture 10_24 and Picture 10_25. Below 30 hours HRT, COD removal is influenced by settling properties (F5/0.6); longer retention times also influence non-settled solids.

I5 = IF (E5 < 30; 1; IF (G5 < 20; (G5 - 10) x 0.39 / 20 + 0.47;
IF(G5<25; (G5 - 20) x 0.14 / 5 + 0.86; IF(G5 < 30; (G5 - 25) x 0.08 / 5 + 1; 1.1))))

The formula refers to Picture 10_17. COD removal by sedimentation (HRT <30 hours) is not influenced by temperature.

J5 = IF (E5 < 24; 1; IF (F17 = 1;1; IF (F17 = 2; 1.08; IF (F17 = 3; 1.12; 1.13))))

A11 = IF (H5 x I5 x J5 <0.98; H5 x I5 x J5; 0.98)

B11 = IF (A11 < 0.5; 1.06; IF (A11 < 0.75; (A11 - 0.5) x 0.065 / 0.25 + 1.06;
IF (A11 < 0.85; 1.125 - (A11 -0.75) x 0.1 / 0.1; 1.025)))

The formula refers to Picture 10_3.

$C11 = A11 / B11$

$D11 = B5 - (C11 \times B5)$

$E11 = C5 - (A11 \times C5)$

$F11 = A5 \times C5 / (A17 + J11)$

$G11 = 75\% \times IF(G5 < 10; 100; IF(G5 < 20; G5 \times 20 - 100; IF(G5 < 25; G5 \times 10 + 100; 350)))$

The formula refers to the rule of thumb by Mara, reflected in Table 22.

$I11 = 0.005 \times IF(H11 < 36; 1 - H11 \times 0.014;$
$IF(H11 < 120; 0.5 - (H11 - 36) \times 0.002; 1/3))$

The formula refers to Picture 10_5.

$J11 = 30 \times A5 \times (C5 - E11) \times I11 \times H11 / 1000$

$A17 = A5 / 24 \times E5$

$C17 = (J11 + A17) / B17$

$E17 = C17 / D17$

$G17 = E17 / F17$

$J17 = A5 \times (B5 - D11) \times 0.35 / 1000 / H17 \times I17$

The formula assumes 350l methane production per kg COD removed.

0 Designing DEWATS

	A	B	C	D	E	F	G	H	I	J
1	General spreadsheet for anaerobic and sedimentation ponds										
2	daily flow	COD in	BOD_5 in	COD/ BOD_5	HRT	settleable SS/COD ratio	ambient temp. °C	BOD_5 removal factors		
3	given	given	given	calcul.	chosen	given	given	calculated acc. to graphs		
4	m^3/day	mg/l	mg/l	ratio	h	mg/l	°C	f-HRT	f-temp	f-number
5	260	2000	850	2.35	72	0.42	25	57%	100%	100 %
6	domestic -> 0.35 - 0.45									
7	treatment data									
8	BOD_5 removal rate	BOD/COD removal	COD removal rate	COD out	BOD_5 out	org. load BOD_5 on total vol.	odourless limit of org. load	desludging interval	sludge accum.	sludge volume
9	calcul.	calcul.	calcul.	calcul.	calcul.	calcul.	calcul.	chosen	calcul.	calcul.
10	%	factor	%	mg/l	mg/l	g/m^3*d	g/m^3*d	months	l/g BOD	m^3
11	57	1.08	53	943	366	171	263	60	0.0023	512
12										
13	dimensions						biogas potential			
14	water volume	depth of pond	total area of pond	width of ponds	total length of pond	number of ponds	length of each pond if equal	methan content	non-dissolv. methane prod.	potential biogas product.
15	calcul.	chosen	required	chosen	calcul.	chosen	calcul.	assumed	assumed	calcul.
16	m^3	m	m^2	m	m	number	m	ratio	ratio	m^3/d
17	780	2.0	646	6.00	107.67	1	107.67	70%	50%	68.67
18										

Table 31:
Spreadsheet for calculating dimensions for anaerobic sedimentation pond (with **short HRT**).
In the example, the pond is extremely long and narrow to facilitate the development of scum in the highly loaded front portion. A baffle wall in the front third supports the effect.
If the pond was squarer, there would be no highly loaded areas, but also no sealing scum layer.
Both options are possible.

	A	B	C	D	E	F	G	H	I	J
1	General spreadsheet for anaerobic and sedimentation ponds									
2	daily flow	COD in	BOD_5 in	COD/BOD_5	HRT	settleable SS/COD ratio	ambient temp. °C	BOD_5 removal factors		
3	given	given	given	calcul.	chosen	given	given	calculated acc. to graphs		
4	m^3/day	mg/l	mg/l	ratio	h	mg/l	°C	f-HRT	f-temp	f-number
5	260	2000	850	2.35	480	0.42	25	92%	100%	108 %
6	domestic -> 0.35 - 0.45									
7	treatment data									
8	BOD_5 removal rate	BOD/COD removal	COD removal rate	COD out	BOD_5 out	org. load BOD_5 on total vol.	odourless limit of org. load	desludging interval	sludge accum.	sludge volume
9	calcul.	calcul.	calcul.	calcul.	calcul.	calcul.	calcul.	chosen	calcul.	calcul.
10	%	factor	%	mg/l	mg/l	g/m^3*d	g/m^3*d	months	l/g BOD	m^3
11	98%	1.03	96	88	17	36	263	60	0.0023	881
12										
13	dimensions							biogas potential		
14	water volume	depth of pond	total area of pond	width of ponds	total length of pond	number of ponds	length of each pond if equal	methane content	non-dissolv. methane prod.	potential biogas product.
15	calcul.	chosen	required	chosen	calcul.	chosen	calcul.	assumed	assumed	calcul.
16	m^3	m	m^2	m	m	number	m	ratio	ratio	m^3/d
17	5.200	2.5	2.432	20.00	212.62	2	60.81	70%	50%	124.29
18										

Table 3
Spreadsheet is the same as Table 3 but is used to calculate dimensions anaerobic-fermentation pond with **long HF**

10.2.10 Aerobic pond

Volume of flow and pollution load are the basic entries for calculation. Key "chosen" parameter is the desired effluent quality (BODout, cell F5). The HRT required to achieve a certain BOD-removal rate depends on the temperature. The curve (Picture 10_29) shows this relationship for a 90% BOD-removal rate.

Picture 10_28 shows how HRT changes with altering treatment performance, defined as BOD-removal rate at 25°C.

Sludge production may be high in aerobic ponds, due to dead algae sinking to the bottom. According to Suwarnarat 1.44g TS can be expected from 1g BOD_5. Assuming a 20% total-solids content in compressed bottom sludge and a 50% reduction of volume due to anaerobic stabilisation, almost 4mm of bottom sludge per gram $BOD_5/m^2 \times d$ organic load would accumulate during one year. At loading rates of 15g $BOD_5/m^2 \times d$, approximately 6cm of sludge is expected per year. Since the surface area plays the major role for dimensioning, the sludge volume has been neglected in the calculation.

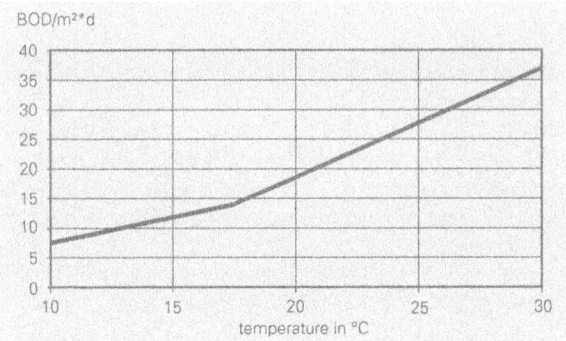

Picture 10_27:
Maximum organic load in relation to temperature for aerobic-facultative oxidation ponds; the influence of sunshine hours has been included

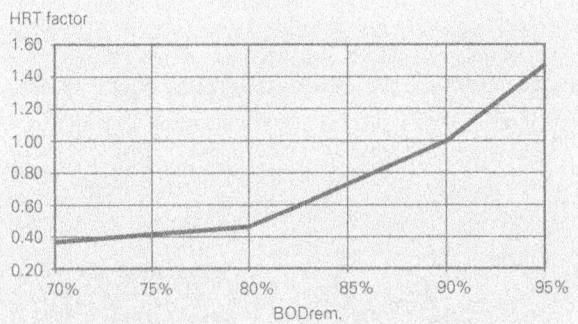

Picture 10_28
Influence of desired BOD removal on HRT in aerobic-facultative ponds based on 25°C

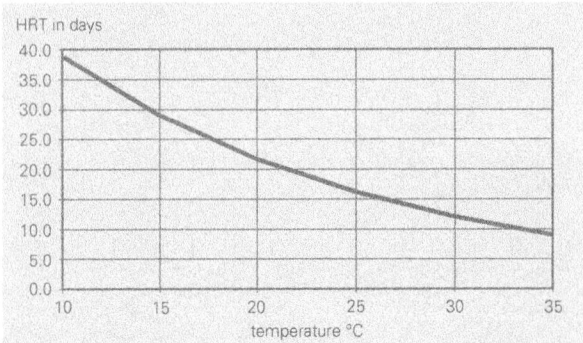

Picture 10_29:
Influence of temperature on BOD removal in aerobic-facultative ponds based on desired BOD removal of 90%

Designing DEWATS

Formulas of spreadsheet for calculation of "aerobic pond"

```
D5 = B5 / C5
G5 = 1 - (F5 / C5)
H5 = G5 x 1 / IF (G5 < 0.5; 1.06; IF (G5 < 0.75; (G5 - 0.5) x 0.065 / 0.25 +1.06;
IF (G5 < 0.85; 1.125 - (G5 - 0.75) x 0.1 / 0.1; 1.025)))
```

The formula refers to Picture 10_3.

```
I5 = B5 - H5 x B5
J5 = IF (G5 < 0.8; (G5 - 0.7) x 0.05 / 0.1 + 0.37; IF (G5 < 0.9; (G5 - 0.8) x 0.54 / 0.1 + 0.46;
(G5 - 0.9) x 0.48 / 0.05 + 1))
```

The formula refers to Picture 10_28.

```
K5 = J5 x IF (E5 < 15; 39 - (E5 - 10) x 10 / 5; IF (E5 < 20; 29 - (E5 - 15) x 7/5;
IF (E5 < 25; 22 - (E5 - 20) x 6 / 5; IF (E5 < 30; 16 - (E5 - 25) x 4 / 5; 12))))
```

The formula refers to Picture 10_29.

```
A11 = 30 x A5 x (C5 - F5) x A12 x L5 / 1000
B11 = IF (E5 < 17; (E5 - 10) x 7.5 / 7.5 + 7.5; (E5 - 17) x 23 / 13 + 14)
```

The formula refers to Picture 10_27.

```
C11 = A5 x C5 / (F11 x G11 x H11)
E11 = IF (IF (F11 = 1; 1; IF (F11 = 2; 1 / 1.1;
IF (F11 = 3; 1 / 1.14; 1 / 1.16))) x (A11 + A5 x K5) / D11 > C5 x A5 / B11; IF (F11 = 1; 1;
IF (F11 = 2; 1 / 1.1; IF (F11 = 3; 1 / 1; 14; 1/1.16))) x (A11 + A5 x K5) / D11; C5 x A5 / B11)
```

The first part of the formula considers the influence of dividing the total pond area into several ponds. The second part compares permitted organic load with calculated HRT.

```
H11 = E11 / F11 / G11
I11 = A5 / D11
K11 = I11 / J11
L11 = I11 + F11 x E11
A12 = 0.0075 x IF (L5 < 36; 1 - L5 x 0.014; IF (L5 < 120; 0.5 - (L5 - 36) x 0.002; 1 / 3))
```

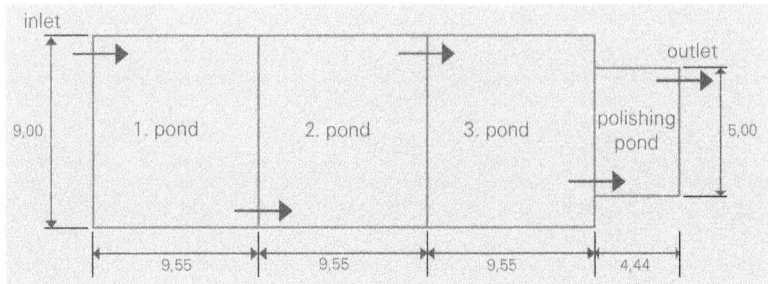

Picture 10_30: Illustration to spreadsheet for the calculation of dimensions o aerobic-facultati ponds

	A	B	C	D	E	F	G	H	I	J	K	L
1	General spreadsheet on aerobic-facultative ponds, input and treatment data											
2	daily flow	COD in	BOD$_5$ in	COD/BOD$_5$	min. water temp.	BOD$_5$ out (wanted)	BOD removal	COD removal	COD out	BOD$_5$ removal factor for HRT	HRT	de-sludging interval
3	given	given	calcul.	calcul.	given	chosen	calcul.	calcul.	calcul.	calcul.	calcul.	chosen
4	m³/d	mg/l	mg/l	mg/l	°C	mg/l	%	%	mg/l	%	days	months
5	20	500	170	2.94	20	30	82	78	108	0.59	12.9	12
6										0.05 - 1.0		
7	dimensions of aerobic-facultative ponds								polishing pond 1 day HRT			total
8	accum. sludge volume	permit org. load BOD$_5$	actual org. load (BOD$_5$)	depth of ponds	total pond area	number of main ponds	width of ponds	length of each ponds	area of polish pond	width of polish pond	length of polish pond	area of all ponds
9	calcul.	calcul.	calcul.	chosen	calcul.	chosen	chosen	calcul.	calcul.	chosen	calcul.	calcul.
10	m³	g/m²·d	g/m²·d	m	m²	No.	m	m	m²	m	m	m²
11	6.3	19.3	13.2	0.9	258	3	9.00	9.55	22	5.00	4.44	796
12	0.00624	l/g BOD		0.9 - 1.2 m								

Table 33: Spreadsheet for calculat dimensions of aerobic-facultative ponds

10.3 Spreadsheets for costings

General background

This chapter helps the reader to produce his or her own tool for calculating annual DEWATS costs. Since economic calculations always incorporate the unknown future, they are never exact. However, it would be reckless to invest in DEWATS without prior economic evaluation. The spreadsheet helps one to calculate annual expenditure, including capital costs, operational costs and maintenance. Expected income from biogas or the sale of sludge for fertiliser may be deducted. To use the spreadsheet, the following data must be collected:
- planning costs, including transport to site and laboratory costs for initial wastewater analysis
- investment costs of buildings, site work and equipment
- assumed maintenance and operating costs
- rate of interest (minus inflation rate)
- wastewater data to calculate possible benefits – and to compare cost per amount of treated wastewater

Formulas of spreadsheet "annual costs of DEWATS"

```
D5 = SUM (A5 : C5)
I5 = SUM (D5 : H5)
J5 = SUM (G9 : K9) + E13 - J13
K5 = SUM (H9 : K9) + E13 - J13
F9 = 1 + E9
G9 = E5 × E9
H9 = (F5 + D5) × (POWER (F9; 20)) × (F9 - 1) / (POWER (F9; 20) -1)
```

This and the following formulas are financial standard operations; the mathematcal expression is:

```
I9 = G5 × (POWER (F9; 10)) × (F9 - 1) / (POWER (F9; 10) -1)
```

The mathematical expression is:

```
J9 = H5 × (POWER (F9; 6)) × (F9 - 1) / (POWER (F9; 6) -1)
```

The mathematical expression is:

```
K9 = SUM (G9 : J9) + E13 - J13
E13 = A13 + B13 + C13 + D13
F13 = A9 × (B9 -D9) × 0.35 × 0.5 / 0.7 / 1000
```

The formula assumes 350l produced biogas per kg COD removed.

```
H13 = F13 × 70% × G13 × 0.85 × 360
J13 = H13 + I13
```

1	Calculating of annual costs of DEWATS									
2	planning and site supervision cost				investment cost					total annual cost	
	A	B	C	D	E	F	G	H	I	J	K
3	salaries for planning and supervision	transport and allowance for visiting or staying at site	cost for wastewater analysis	total planning cost includ. overheads and acquisition	cost of plot incl. site preparation	main structures of 20 years' durability	secondary structures of 10 years' durability	equipment and parts of 6 years' durability	total investment cost (incl. land and planning)	total annual cost (including land)	total annual cost (excluding land)
4	l.c.	l.c.	l.c.	l.c.	l.c.	l.c.	l.c.	l.c.	l.c.	l.c.	l.c.
5	1,200	650	500	2,350	150,000	295,000	9,000	3,000	459,350	74,359	62,359
6	wastewater data				annual capital costs						
7	daily wastewater flow	strength of wastewater inflow	COD/BOD ratio of inflow	strength of wastewater outflow	rate of interest in % p.a. (bank rate minus inflation)	interest factor $q=1+i$	on investment for land	on main structures of 20 years' lifetime (incl. planning fees)	on secondary structures of 10 years' lifetime	on equipment of 6 years' lifetime	total capital costs
8	m^3/d	mg/l COD	mg/l	mg/l COD	%		l.c./year	l.c./year	l.c./year	l.c./year	l.c./year
9	20	3,000	2	450	8%	1.08	12,000	30,286	1,341	649	37,179
10	operational cost				income from biogas and other sources						explanat.
11	cost of personal for operation, mainten. and repair	cost of material for operation, mainten. and repair	cost of power (e.g. cost for pumping)	cost of treatment additives (e.g. chlorine)	total operational cost	daily biogas production (70% CH_4, 50% dissolved)	price 1 litre of kerosene (1m^3 CH_4=0.85 l kerosene)	annual income from biogas p.a.	other annual income or savings (e.g. fertiliser, feeds)	total income per annum	l.c. = local currency; mg/l = g/m^3
12	l.c./year	l.c./year	l.c./year	l.c./year	l.c./year	m^3/d	l.c./litre	l.c./year	l.c./year	l.c./year	
13	100	100	50	0	250	12.75	2.69	7,347	0	7,347	

Table 34:
Spreadsheet for the economic calculation of DEWATS (based on annual costs).

Viability of using biogas

Whether using biogas is economically viable depends on the additional investments to facilitate storage, transport and utilisation of biogas – and, if these costs can be recovered by the income generated by biogas production within a reasonable time. The payback period is considered to be an adequate indicator of viability.

Formulas of spreadsheet "viability of biogas"

$B4 = 6.5\% \times A4$

For rough calculation, it is assumed that additional construction costs are 6.5% of original costs, which includes investment for making the reactor roof gas-tight, providing additional volume for gas storage, and for gas distribution and supply pipes.

$D4 = 50\% \times C4$

To guarantee permanent gas supply, additional care must be taken at the site; it is assumed that operational costs are 50% higher than without biogas use.

$F4 = B4 / (E4 - D4)$

Negative values show that costs will never be recovered.

	A	B	C	D	E	F
1	Economic viability of using biogas					
2	investment cost without use of biogas	additional constr. cost to facilitate use of biogas	operational cost without use of biogas	additional operational cost to use biogas	income from biogas	pay back period of additional cost
3	l.c.	l.c.	l.c./year	l.c./year	l.c./year	years
4	307,000	19,955	250	125	3,650	5.7

Table 35: Spreadsheet for calculating the viability of necessary measures to facilitate biogas utilisation

10.4 Using spreadsheets without a computer

Not everybody uses a computer; some may not even have access to one. But, computer formulas may also be useful to those who usually work with a pocket calculator. The following explanations are presented for them. The calculation for the septic tank (see Table 36) is a good example:

A computer table is made up of cells. The location of each cell within the table is described by columns A.....X, AA...AX, etc. and rows 1.....>1000. Each cell within the table, therefore, has an exact address. For example, the first cell in the top left corner has the address A1 (column A, row 1). In the table below, cell J10 reads m³/d and cell D5 reads 633. Cell I11 reads 23.25; this figure is the result of a formula hidden "under" it. On the computer, the formula appears in the headline every time one clicks on the cell. These formulas can also be applied without a computer, in connection with the various graphs. One must realise, however, that the computer writing differs from normal mathematical writing in some points: for example, 4/(3×2) is written as = 4/3/2 on the computer, and 4×2/3 may be written either 4*2/3 or 4/3*2.

	A	B	C	D	E	F	G	H	I	J
1	General spreadsheet for septic tank, input and treatment data									
2	daily waste- water flow	time of most waste- water flow	max. flow at peak hours	COD inflow	BOD$_5$ inflow	HRT inside tank	settleable SS/COD ratio	COD removal rate	COD outflow	BOD$_5$ outflow
3	given	given	calcul.	given	given	chosen	given	calcul.	calcul.	calcul.
4	m³/day	h	m³/h	mg/l	mg/l	h	mg/l	%	mg/l	mg/l
5	13.0	12	1.08	633	333	18	0.42	35	411	209
6				COD/BOD$_5$ ->	1.90	12 - 24	0.35 - 0.45 domestic		BOD rem. ->	1.06
7	dimensions of septic tank									
8	deslud- ging interval	inner width of septic tank	min. water depth at outlet point	inner length of first chamber		length of second chamber		volume incl. sludge	actual volume of septic tank	biogas 70% CH$_4$ 50% dis- solved
9	chosen	chosen	chosen	requir.	chosen	requir.	chosen	requir.	check	calcul.
10	months	m	m	m	m	m	m	m³	m³	m³/d
11	12	2.50	2.00	3.13	3.10	1.56	1.55	23.46	23.25	0.72
12							sludge l/g BOD rem.	0.0042		

Table 36:
Sample spreadsheet used to help understand computer formulas

Cell A5 and all other bold written figures contain information to be collected and do not comprise formulas. The cells with hidden formulas are these:

$C5 = A5 / B5$

Meaning: 13.0 [m³/d] / 12 [hours] = 1.08 [m³/hours]

$H5 = G5 / 0.6 \times IF (F5 < 1; F5 \times 0.3; IF (F5 < 3; (F5 - 1) \times 0.1 / 2 + 0.3;$
$IF (F5 < 30; (F5 - 3) \times 0.15 / 27 + 0.4; 0.55)))$

Meaning: (0.42 [mg/l / mg/l] / 0.6 [a given factor found by experience]) multiplied by the value taken from Picture 10_6 at 18 hours HRT (shown in cell F5).

The calculation is, therefore:

$(0.42 / 0.6) \times 0.495 = 0.35 = 35\%$ (which is shown in cell H5)

$I5 = (1 - H5) \times D5$
$(1 - 0.35) \times 633 = 411$ (shown in cell I5)

$J5 = (1 - H5 \times J6) \times E5$
$(1 - 0.35 \times 1.06) \times 333$

$E6 = D5 / E5$
$633 / 333 = 1.90$

$J6 = IF (H5 < 0.5; 1.06; IF (H5 < 0.75; (H5 - 0.5) \times 0.065 / 0.25 + 1.06;$
$IF (H5 < 0.85; 1.125 - (H5 - 0.75) \times 0.1 / 0.1; 1.025)))$

This formula refers to Picture 10_3. Since cell H5 (the removal rate) is 35%, the value of cell J6 is found in the graph and equals 1.06.

$$D11 = 2/3 \times H11 / B11 / C11$$
$$((2/3) \times 23.46) / (2.50 \times 2.00) = 3.13$$

$$F11 = D11 / 2$$
$$3.13 / 2 = 1.56$$

$$H11 = IF (H12 \times (E5 - J5) / 1000 \times A11 \times 30 \times A5 + C5 \times F5 < 2 \times A5 \times F5 / 24; 2 \times A5 \times F5 / 24;$$
$$H12 \times (E5 - J5) / 1000 \times A11 \times 30 \times A5 + C5 \times F5) + 0.2 \times B11 \times E11$$

The formula refers via cell H12 to Picture 10_5; cell H12 must be calculated first. The formula H11 states that the total volume must be at least twice the sludge volume. One has to check whether the total volume must be calculated via the hydraulic retention time or via the double sludge volume.

The total volume equals the sludge volume, which is $0.0042 \times (333 - 209) \times 12 \times 30$ [days/month] $\times$ 13.0 / 1000, plus the volume of water, which is $1.08 \times 18 = 21.88 m^3$. This is compared to $2 \times 13.0 \times 18 / 24$ [hours/day], which equals $19.50 \ m^3$. Since 21.88 is the larger of the two, it must be used. Finally, the volume of 20cm of scum must be added, which is $0.2 \times 2.50 \times 3.10 = 1.55$. The total volume is $21.88 + 1.55 = 23.43$ (the computer is slightly more exact and states $23.46 m^3$ in cell H11.

$$I11 = (E11 + G11) \times C11 \times B11$$
$$(3.10 + 1.55) \times 2.00 \times 2.50 = 23.25 m^3$$

$$J11 = (D5 - I5) \times A5 \times 0.35 / 1000 / 0.7 \times 0.5$$
$$(633 - 411) \times 13.0 \times 0.35 \times 0.5 / (1000 \times 0.7) = 0.72 m^3$$

$$H12 = 0.005 \times IF (A11 < 36; 1 - A11 \times 0.014; IF (A11 < 120; 0.5 - (A11 - 36) \times 0.002; 1/3))$$

The last formula refers to Picture 10_5. The desludging interval is 12 months (cell A11), which results in a value of approximately 80% in the graph; this figure is multiplied by the sludge-production figure of 0.005.

The calculation is, therefore:

$$0.8 \times 0.005 = 0.004 \text{ (the computer calculates 0.0042).}$$

1 Project components: sanitation and wastewater treatment – technical options

The other components of DEWATS and DEWATS/CBS systems along the sanitation chain before and after the wastewater treatment are:
- toilets
- collection systems
- reuse and disposal systems, including sludge treatment and biogas applications
- construction management
- management of operation & maintenance
- health and hygiene behaviour

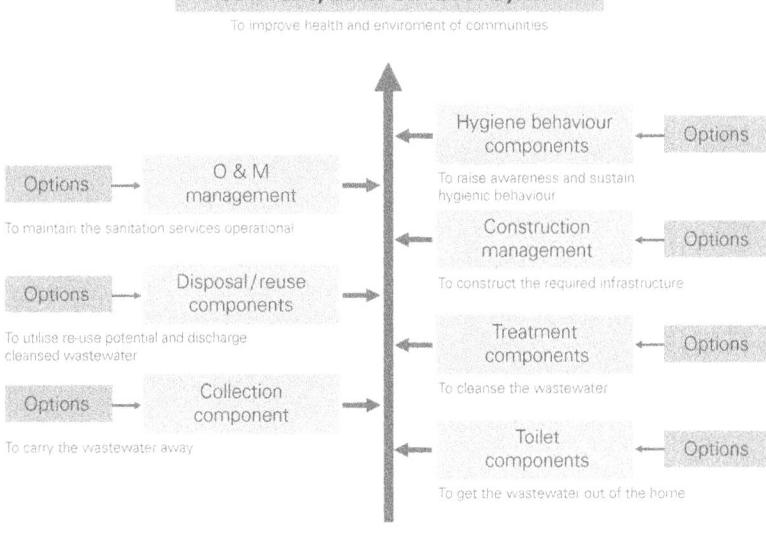

Figure 11_1
Community-Based Sanitation System: technical options along the sanitation chain

Each component presents a wide range of possible technical options. To select the most appropriate solution for a location, the options must be assessed with the help of various criteria, such as capacity, costs, self-help compatibility, operation & maintenance, replication potential, reliability, convenience and efficiency.
While operational and process related issues are dealt with in chapters 5 and 6, this chapter presents technical options for toilets, collection systems, sludge accumulation and treatment, the reuse of wastewater and sludge as well as biogas utilisation.

11.1 Toilets

When communities use hygiene and sanitation methods that fit their real needs and abilities, they will enjoy better health. In most cases, the toilet component is the users' prime concern. There are many reasons why users might prefer one sanitation option over another, beside, health, better water supplies or improved hygiene:
- Privacy – the need for privacy makes it important for a toilet to have a good shelter. Providing a door or enclosed entrance, or constructing it away from busy locations, makes the toilet nicer to use
- Safety – a poorly constructed toilet can be dangerous to use. If it is far from the home, women may be in danger of sexual violence. A toilet must be well-built and in a safe location
- Comfort – people prefer to use a toilet with a comfortable place to sit or squat, and a shelter large enough to stand up and move around in.
  Children, the elderly or people with disabilities have special needs to permit comfortable use
- Cleanliness – no one wants to use a dirty and smelly toilet. Toilet areas should be well-lit and ventilated. Easy-to-clean surfaces and cleary defined of cleaning responsibilities help to ensure that toilets are well-kept
- Respect – a well-kept toilet brings status and respect to its owner; this may be an important reason for people to spend money and effort to build one

The following section describes a selection of possible toilets – from common, hazardous models to recommended options. No one toilet design is right for every community or household. It is important, therefore, to understand the benefits and risks of each and to adapt designs to suit local conditions and cultural preferences.

11 Project components: sanitation and wastewater treatment – technical options

11.1.1 Common practices to be discouraged

Open defecation

The lacking of sanitation facilities forces large parts of the world's population to defecate openly. Depending on the location, refuge is sought in the forest, jungle, lakes, rivers or the ocean. Apart from lacking privacy and the obvious associated hygienic-health risks, open defecation places humans in a vulnerable situation. Women and children can easily become targets of sexual abuse or violence. In many cases, parents also worry about the safety of their children, because of poisonous snakes or other potential dangers in the bush or jungle.

Figure 11_2 and 11_3: Residents returning from distant open-defecation areas; a bush toilet

Overhang latrine

Overhung latrines are usually built from bamboo or wood and sited above the surface of water bodies (such as rivers, ponds or lakes). Excreta fall directly into the water, where they are decomposed. Usually it is a public facility, which serves an entire or part of a community. This type of latrine pollutes the receiving water body, which can no longer be used as a fresh-water source (exceptions may include very rural settings with large or fast-moving water bodies). Furthermore, the system is usually inconvenient, as it is located away from settlements.
The exposed location affords users with little privacy.

Picture 11_4:
Overhang latrin

11.1.2 Closed pit toilets

Closed pit toilets are very common in developing countries and are always located outside the house.

They consist of a deep pit, which is covered by a platform with a shelter. The platform has a hole in it and a lid to cover the hole when it is not in use. The platform can be made of wood, concrete, or logs covered with earth. Concrete platforms help to keep water out of the pit and are very durable. A closed pit toilet should have a lining or concrete-ring beam to prevent the platform or the pit itself from collapsing. The average pit depth of 3m is usually limited by the groundwater table or rocky underground. The underground of the latrine should be water pervious. Dry anal cleansing is advantageous to minimise water content. No sullage treatment is included.

The latrine can be used until it is filled up to half a metre below the top; its lifetime depends on the number of users and pit size. At that point, space is required for emptying for the pit – which is to be discouraged for hygienic reasons – or relocation of the toilet.

To prevent groundwater pollution and increased health risks, pit toilets are only suitable in flood-free areas, where the highest seasonal groundwater table lies well below the floor of the pit. The system has a large potential for odour, insects and hygiene hazards, especially if not cleaned regularly.

Ventilated improved pit toilets (VIP)

The VIP toilet is a kind of closed pit toilet that reduces smells and flies.
The design and applicability is the same as for a normal pit latrine – made of a latrine superstructure, a pit-cover slab and a lid-covered hole for defecation. The only difference is the ventilation pipe, provided with a durable fly-screen on the top and reaching high above neighbouring roof-tops. A dark-coloured ventilation pipe should be chosen, to promote convection, or upwards air-flow within the pipe. A disadvantage of VIP latrines is that the toilet room must be kept relatively dark to encourage flies to travel towards the light at the end of the ventilation pipe, where they are trapped and die at the fly-screen. Good maintenance of the screen is important to ensure convenience and healthy conditions. Dry anal cleansing is advantageous to minimise water content. No sullage treatment is included. It is common to relocation the latrine after the pit is full.

Shallow (composting) pit toilets for tree planting

The design is similar to that of a VIP latrine – made of a latrine superstructure, a pit cover slab with ventilation pipe and a lid-covered hole for defecation. The system is better at reducing the risk of groundwater pollution when compared to other closed pit toilets because the pit is very shallow (maximum depth of 0.5 to 1m). It thereby ensures that the faecal matter is contained within the biologically active upper soil zone, where it can be decomposed.

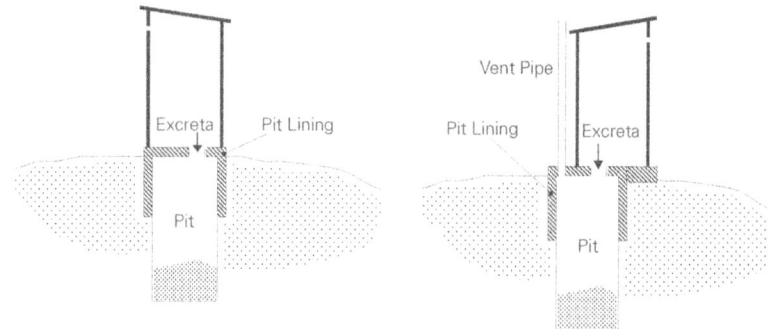

Picture 11_5: Standard pit and VIP latrine

1 Project components: sanitation and wastewater treatment – technical options

When the pit fills, the toilet house, including the concrete slab, is moved to a neighbouring location and a tree is planted on the site of the first pit. Shallow pit toilets are most appropriate where there is space and people want to plant trees. They can be constructed in locations where rocky underground prevents the digging of deeper pits.

The design can be improved by installing a urine separation pan with a collection container, in sandy-soil conditions, to avoid nitrogen infiltration. The system is not suitable in areas with a rocky surface, extremely high groundwater or flooding.

 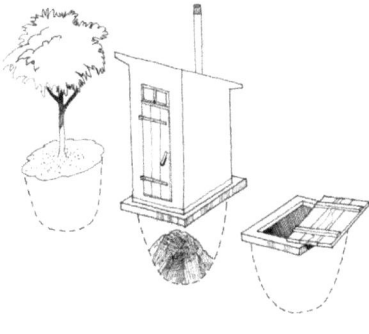

ture 11_6 to 11_9:
Shallow (composting) pit toilets for tree planting.
Source: Stockholm Environment Institute, 2004

11.1.3 Composting toilets

Composting toilets retain faeces and urine and turn them into soil conditioner and fertiliser. Sitting or squatting models are available.

The composting latrine consists of a squatting plate, which is placed over a watertight vault usually constructed above the soil. The vault is ventilated through a pipe, which extends above the surrounding rooftops and has a fly-screen at the top. To support the composting process it is necessary to add dry organic material, such as straw, leaves, sawdust, soil or vegetable waste, at daily intervals. This reduces smells and helps the waste to break down. Different techniques can be applied to reduce the water content, thus guaranteeing optimal aerobic conditions. Under the right conditions, the mix will heat up, thereby killing most germs, including roundworm eggs (the hardest to kill). After sufficient treatment time (usually one year), the composted material is removed for use as a fertiliser. To be safe, it is best to mix it into a compost pile, where it will break down more. Then it can be mixed into the soil for planting.

Due to the importance of the moisture content in the chamber, composting latrines are only suitable for communities using dry cleansing material or with separate wash-water drainage and treatment. Since the water content within the vault must be monitored, the users must fully understand and appreciate the process to ensure proper operation of the system without odour or insect nuisance.
The toilet is normally located outside the house and can be used for many years, if operated properly. The system is convenient in rural areas where composting is traditionally practised. No sullage treatment is included.

A variation of the system includes two vaults, which are alternately in use. While one vault is being used, the content of the other is topped up with 30cm of soil and covered with a concrete slab. With time, the contents are dehydrated through evaporation and decomposed by micro-organisms. When the second pit is full, the odourless and partially disinfected compost can be removed from the first pit. If it is still wet and smells, further composting or storage in a dry place is advised. Wear gloves, and wash hands after handling the fresh fertiliser.

11 Project components: sanitation and wastewater treatment – technical options

11.1.4 Dry, urine-diversion toilets

Dry, urine-diversion toilets combine toilet house and treatment facility into one above-ground structure. They can be located inside the house, attached to it or left as a free-standing unit in the yard. Urine and faeces are collected separately by special toilet models of various designs. Sitting and squatting models are available.

The super-structure is elevated to create sufficiently sized faeces-storage volume below the cover slab. These storage chambers are waterproofed to ensure dry conditions, even in the case of heavy rain or flooding.

The key to successful operation is the fast dehydration of the faeces. Laying bamboo, cornstalks, branches or other dry plant matter on the floor of the chamber before initial use facilitates the drying process. Furthermore, a handful of ashes, sawdust or dry soil sprinkled over the faeces after defecation will to absorb moisture and avoid fly breeding.

 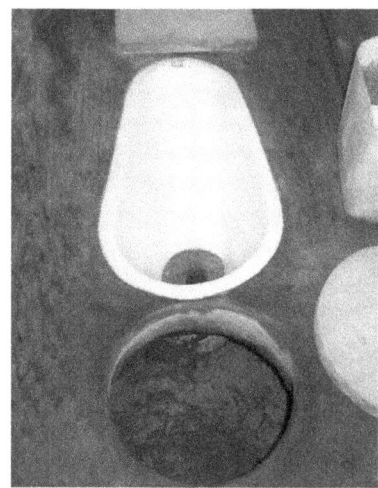

Figure 11_10 and 11_11: Selection of urine-diversion and, squatting toilet models

A ventilation pipe, which extends above the surrounding rooftops and has a fly-screen at the top, causes a constant draft into the toilet, thereby drying the faeces and avoiding smell. Ventilation is increased by using black chamber-access doors facing the sun.

From the separation toilets, urine can be led to collection containers. If collected, it should be treated by air-tight storage for three to six months before being diluted 10 to 1 with water and used as a liquid fertiliser rich in phosphorous and nitrogen. Alternatively, urine – together with water used for anal cleansing – can be led into an evapo-transpiration reed-bed next to the toilet house. Its plants are cut back periodically, chopped into small pieces and added to the processing vault after drying. Good experiences with the system have been reported in South India, even in humid conditions. The traditional Vietnamese double-vault toilet works in the same way but only in combination with dry anal cleansing and urine utilisation for agriculture purposes.

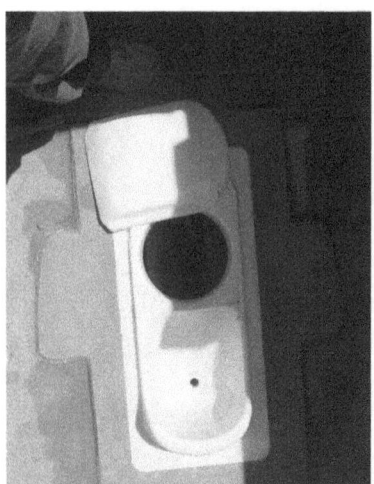
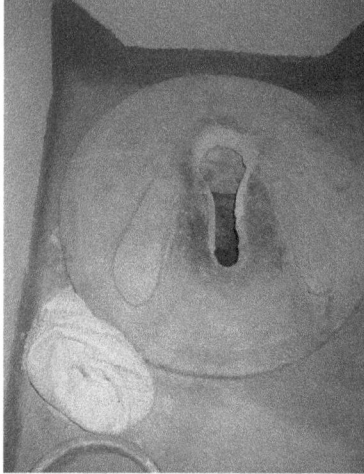

Picture 11_12 and 1
Selection of urin
diversion and,
squatting toilet
models

Faecal storage and treatment can be practised with two possible systems:

a) Two-chamber system: the compartment below the toilet is divided into two chambers. When one chamber is full, it is closed and the second one is used. When the second is full, the first is emptied. The toilet model either has two faecal openings (one leading to each chamber), or the toilet bowl can be removed and turned around to use the other chamber.
b) Storage receptacles: the compartment below the toilet contains several containers. Plastic bins or locally produced reed-baskets can be used. When one of the containers fills, the chamber is accessed and the full container is replaced with an empty one. The full container remains in the compartment. When all storage capacity has been exhausted, the full container with the longest storage time is removed and emptied. Reed-baskets are perfect if further composting is desired.

Access to the faeces chamber can be through a water-tight door, a concrete slab or a temporary hole (weak mortar brickwork) in the chamber wall.

The system is suitable for all geographical conditions – particularly in regions with water scarcity, high groundwater table, flooding or rocky soil. Implementation requires the users to have intensive training. It is not recommended for public or communal toilets, as there is a high risk of misuse.

11.1.5 Pour-flush toilets

Pour-flush toilets are very common; sitting and squatting models are available. Excreta are washed away with approximately 0.5 to 2 litres of water poured into the pan with a scoop. These toilets should only be applied, therefore, where adequate amounts of flush water are available. Since they have a water seal against odours and insects, pour flush-toilets can be located within the house, if desired. Where water is required for anal cleansing, pour-flush toilets are particularly suitable because the same water can be used for flushing. As no complex mechanical devices are needed for operation, the toilets are robust and rarely require repair. Since water is available near and in the toilet, cleaning is very easy.

Pour-flush toilets use a plastic, fibreglass, or cement bowl or squatting pan set into a concrete platform. The concrete platform can either be placed directly over a pit, or it can be connected by pipe to one or two pits. Alternatively, the pipes can feed into a wastewater-collection system or directly into other treatment units (i.e. septic tank).

Pour-flush toilets with one leach pit

Single leach pits are made of a latrine superstructure and a WC pan with a water seal. A collection pipe, 100mm in diameter, is laid at a gradient of at least 1 in 20, if the pit is off-set. The wastewater is discharged into a pit lined with water-pervious brick or stone work. Pits should be covered with reinforced-concrete slabs, stone slabs or wooden planks, secured against mischief by children.

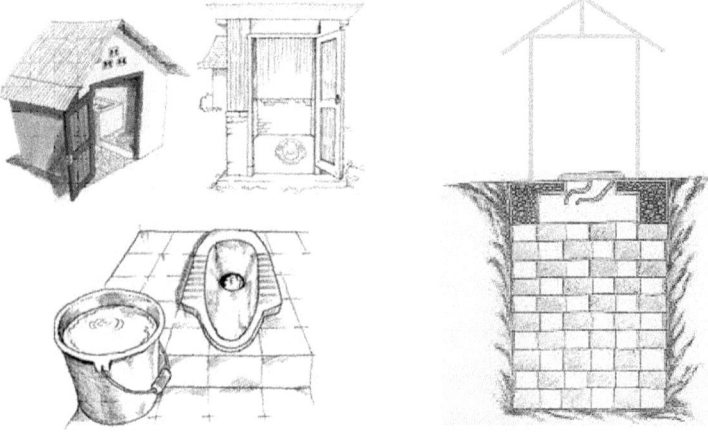

Picture 11_14:
Pour-flush toilet

Picture 11_15:
Pour-flush toilet
with one leach pit

1 Project components: sanitation and wastewater treatment – technical options

Pour-flush toilet with single leach pit

One-pit pour flush toilets can only be used until it the pit is full. A five-headed family will fill a lined pit of two metre depth and 1 meter diameter in approximately 5 years and emptying is required before continued use is possible. Desludging should be provided by professional service providers to minimise health risks. It is easier if the pit is off-set and not directly under the superstructure. Pour flush pit toilets should be applied only in flood-free areas, where the highest seasonal groundwater table lies at least 3m below ground level.

Pour-flush toilet with two leach pits

When there are two pits, a valve directs the wastewater to the pit currently in use. The first pit is used until it is nearly full. Then waste is diverted into the second pit. Soil is added to the first pit and its contents are left to settle for at least two years, then it can be emptied without any great risk of illness from germs.

For a family of five, two pits measuring one metre deep and one metre in diameter would need alternating approximately every three years. The distance between the pits should be at least the same as the depth of the pits. Pour-flush pit toilets are only appropriate in flood-free areas, where the highest seasonal groundwater table is more than 3m below ground level.

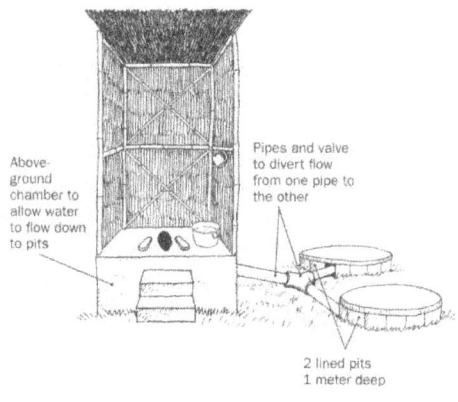

Picture 11_16: Pour flush toilet with two leach pits (toilet house shown without door)

Pour-flush toilet with individual septic tank and French-drain gravel filter

Pour-flush toilets can also lead the wastewater into a small on-site treatment facility. Septic tanks are watertight containers, which provide primary treatment by separating, retaining and partially digesting settleable and floatable solids in wastewater. Septic-tank effluent must receive proper secondary treatment before being discharged to the groundwater or surface water bodies. Directly ensuing soakage pits should not be applied, if the vertical distance from the bottom of the soakage pit to the highest seasonal groundwater is less than 1.5 metres. In these cases, septic tanks can be combined with French-drain filters or equivalent treatment. Septic tanks accumulate sludge which must be emptied after approximately five years and treated separately.

French-drain filters are simplified horizontal, gravel filters for on-site sanitation where there are space constraints and a high groundwater table. They provide simple filtration and anaerobic treatment, where high groundwater tables prevent direct septic-tank effluent infiltration. At the end of the French-drain filter, water is infiltrated to the soil though a plant-bed.

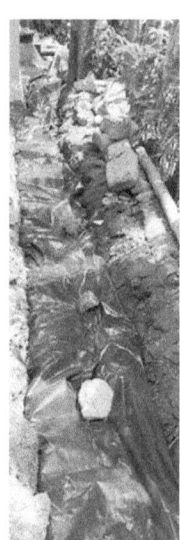

Picture 11_17 to 11_
Construction of a
French-drain filte
connecting a pla
septic tank with
plant-bed

11 Project components: sanitation and wastewater treatment – technical options

Pour-flush toilet attached to wastewater-collection system

Since pour-flush toilets operate with water, the waste can be washed into a local wastewater-collection system, which transfers the excreta to a centralised or decentralised-treatment facility. For more details on wastewater-collection systems, please refer to section 11.2.

11.1.6 Community toilet blocks

Community toilet blocks usually consist of a number of toilet compartments. A large variety of available superstructure options can also include bathrooms, public water-points and laundry facilities.

Each toilet should not be shared by more than six households or 25 people. Integrated concepts can include treatment options such as septic tanks or baffled reactors. Community toilets are a suitable CBS option in settlements where the majority of the households don't have toilets. For convenience, communal toilet blocks should be no further than 50 metres walk.

Past experience has shown that maintaining and operating community toilets properly is a major obstacle for their sustainability. User fees are a "must" to finance routine operation and maintenance services, which ought to be carried out by permanent or part-time O & M staff employed by community groups or private-service providers.

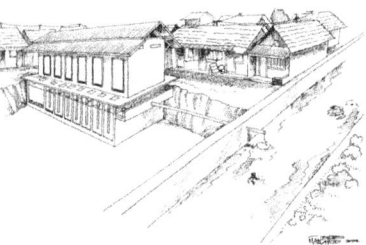

Figure 11_21 and 11_22: Community toilet blocks

11.2 Collection systems

11.2.1 Rainwater drains

Systems with open ditches for discharging rainwater are quite common in the urban areas of developing countries. The ditches usually drain rainwater into rivers or, sometimes, into agricultural-irrigation canals. The unauthorised discharge of domestic waste or drainage of sullage through such a system is a health hazard and should be discouraged.

Covered rainwater drains

Covered rainwater drains are often used to collect wastewater in areas which lack conventional sewerage systems. Drains are covered by concrete slabs to stop them being blocked up by litter and to prevent people from coming in contact with their contents. So that rainwater can enter the system, periodic inlets in the drain covers are required. Theoretically, connected treatment plants would have to be designed for the purification of combined flows – rainwater and domestic wastewater – which requires a very high treatment capacity and investment. Such systems present a temporary solution, where no other system of wastewater collection is available, but it should be replaced by an improved system as soon as possible. The system smells, promotes insect breeding – and remains a health hazard.

Picture 11_23:
Open and closed rainwater drains

1 Project components: sanitation and wastewater treatment – technical options

11.2.2 Conventional gravity sewerage

In conventional gravity sewerage, domestic wastewater flows to a treatment facility via a system of concrete pipes. The system consists of house connections, which lead to a reticulation sewer line, normally laid beneath the main roads. There are inspection manholes every 70m along the route.

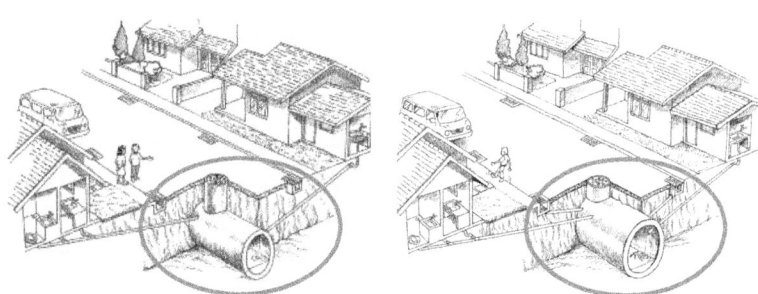

Figure 11_24 and 11_25: Conventional gravity sewerage systems:
11_24: no stormwater connection to main sewer
11_25: stormwater connected to (bigger) main sewer

So that the system can be cleaned, the minimum diameter is usually 200mm (D). To avoid solids deposit, minimum velocity of 0.5m/s is required. The maximum velocity should not exceed 6 to 8m/s. The necessary gradient of the pipes is, in part influenced by their diameter. In preliminary design, the gradient (I_S) can be estimated through the equation $I_S=1/D$. In flat areas, conventional sewer systems can demand very deep and expensive excavation. To avoid excessively deep sewers in large systems, it is necessary to use either a flushing tank or construct a pumping station. In Europe, pipes are usually laid at a minimum depth of 1.5 to 2.0m to guarantee load rating suitable for normal traffic as well as frost protection.

The maintenance of the reticulation system plus the operation and maintenance of possible pumping stations make up the operating costs.

Combined gravity sewerage

In combined gravity sewerage, domestic wastewater flows to a treatment facility together with collected rain- or stormwater, in a similar system to the conventional gravity sewerage. However, since the system must be designed to handle peak flow, much bigger pipe diameters are required for the mixed flow; diameters in the range of 300 to 1,200mm are common. Furthermore, inlets for rainwater from roof and street run-off are necessary.

Just because such gravity systems are currently considered the standard solution in most developed countries, does not mean that the conventional or the combined sewerage system is the optimal solution under all conditions. Engineers should compare all feasible options on an economic and technical basis.

Seperated gravity sewerage

As shown in Picture 11_24, stormwater is not collected together with domestic wastewater but drained seperately. This is today's preferred solution. Wastewater-treatment systems are prevented from stormwater shock loads. The advantages are twofold:
- the biology of the treatment system will be kept stable and does not have to adapt to different concentrations of wastewater (dilution)
- the wastewater-treatment system does not have to be oversized in terms of treatment volume (due to hydraulic peak loads)

11.2.3   Simplified gravity sewerage

Simplified gravity-sewerage systems function like their conventional, larger counterparts. But the design criteria for construction have been simplified so that they just comply with minimum hydraulic requirements. As a result, the pipes made from plastic or concrete have smaller diameters and are usually laid at a flatter gradient and a shallower depth. The system can also cope with fewer inspection manholes. Although the costs are reduced, there is an increased probability of malfunction, resulting in more intensive operation and maintenance work.

1 Project components: sanitation and wastewater treatment – technical options

Condominial gravity sewerage

Condominial sewerage is usually based on a PVC-piping system with a minimum diameter of 100mm, leading wastewater towards a nearby treatment facility or towards another sewer network. Pipes are laid at a flat gradient and routed through private land, such as frontyards, backyards (in-block) or pavements. So the required tyre-load capacity is considerably less than for in-road systems. Consequently, it is possible to lay the pipes at a shallow depth. Backyard and frontyard systems require a minimum cover of 20cm, while cover under pavement should be 40cm. Another advantage of backyard sewers is the reduced piping length, resulting in reduced costs. Furthermore, shallow condominial sewerage systems do not require large, expensive manholes.

Simple inspection chambers (located every 20m) and junction boxes at sewer connection points are usually sufficient. As with all systems, who's responsible for maintenance should be clearly defined.

Figure 11_26 and 11_27: Backyard and frontyard condominial gravity sewerage

Small-bore sewerage

Small-bore systems, also called "solid-free sewers", "common effluent drains" or "settled sewerage", receive the effluent from individual or shared household septic tanks. Hence, coarse solids are removed and only the liquid part of sewage enters the sewerage system. Unlike conventional gravity systems, no self-cleansing flow-velocity is required. As a result, small-bore sewers can be operated with less water, allowing the connection of (low) flush toilets (including pour-flush) from households served by a standpipe or yard tap. The pipes have smaller diameters. Flow is driven by the elevation difference between inlet and outlet, and, therefore, can be installed very close to the surface in all types of terrain and even allow inflective gradients.

Simplified sewerage systems, the clogging and blocking of pipes is very unlikely, because of the pre-treatment in septic tanks. This effectively reduces the amount of maintenance needed on the piping system, although regular septic-tank emptying is essential.

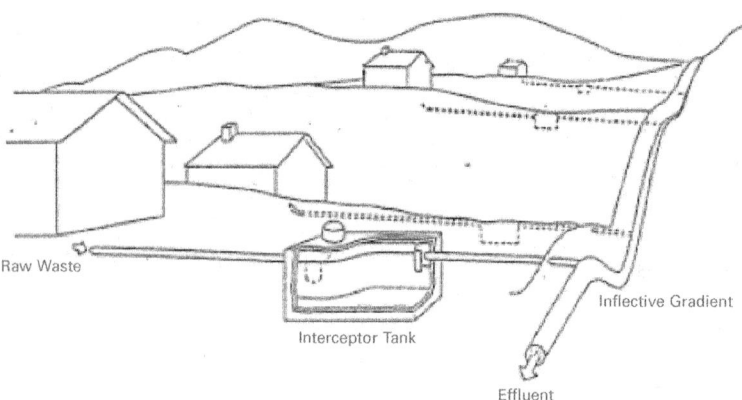

Picture 11_28:
Small-bore
sewerage system

A 100mm-diameter pipe at a slope of at least 1:60 is required to connect the toilet to the septic tank. The level of the tank should not be deeper than necessary, so that the maximum potential energy (arising from its elevation) is available for the flow in the main sewer. At the least the first two metres of the connecting pipe from the septic tank to the plot boundary should have a diameter slightly smaller (50mm diameter) than the sewer main. This reduces the risk of blockage in the main sewer. Any misuse of the tank would then result in the plot-owner being inconvenienced rather than the whole neighbourhood.

The small-bore sewer mains should consist of plastic pipe with a minimum diameter of 100mm, installed at a depth of at least 300mm on plots, 1m on public lands or roads, and 1.2m when crossing roads. Clean-out points should be located at the upstream ends of the system, at the intersection of sewer lines, at major changes of direction, high points, and intervals of 150 to 200m in long, flat sections. These provide access to the sewer inspection and flushing during sewer cleaning. Manholes are not required.

Unlike conventional gravity sewers, small-bore sewers can alternate between open channel and pressure flow, taking maximum advantage of the elevation difference between the upstream and downstream ends of the sewer. Care must be taken that the hydraulic grade line during peak flow does not rise above the invert of the septic-tank outlets. If this is assured, the sewer may have low points or dips and can curve to avoid objects. High points of the sewer should be ventilated. As the sewer is not intended to carry solids, it is designed on hydraulic considerations only.

Pumping stations are only required where elevation differences do not permit gravity flow. If this is the case, permanent electricity supply and professional maintenance services are required for sustainable operation.

Because of the nature of the effluent from the septic tanks, the effluent of small-bore sewers is highly corrosive and odorous. If required, pumps and pump wells should be protected against corrosion and odour emission.

11.2.4 Vacuum sewerage

Vacuum wastewater-collection systems save water by using air as the main transport medium within the pipelines, by maintaining a low pressure of 0.6bar within the sewer network with vacuum pumps. The sewerage lines can be installed very close to the surface in all types of terrain – and can even transport wastewater around obstacles and up-hill. They require a power supply at one centralised location. The system consists of three basic elements: collection chambers, sewer network and a vacuum station.

Any type of (low-)flush toilet (including pour-flush) can be used. The wastewater drains from the household to a collection chamber by gravity. These chambers are not mechanised and can be located on or near the plot, and can receive wastewater streams from several neighbouring households. When the wastewater in the collection chamber reaches a certain level, an interface valve is triggered and opens automatically without an external power supply. This valve connects the collection chamber to the low-pressure sewer network. Together with the wastewater, about six times more air will be sucked into the system. The air is used as a transport medium for the wastewater, reaching transport velocities of 4 to 6m/s on the way to the vacuum vessel or pump sump in the vacuum station. When the collection chamber is emptied, the interface valve closes again. The pump sump is connected to a treatment facility.

Collection chambers must be made of watertight, smooth, corrosion-resistant material and big enough to take 25% of the average daily flow. The pipe network is made of PE-HD (polyethylene, high density) or PVC (polyvinyl chloride); both can be electro-welded or solvent-welded (cemented). Only the short gravity sewer from the house to the collection chamber must have a minimum diameter of 100mm and be laid at 1:60 or steeper.

1 Project components: sanitation and wastewater treatment – technical options

The minimum size of the vacuum sewer grid should be 90mm diameter. Pipelines should be designed to withstand the internal suction pressure and temperature. The minimum pressure rating of selected pipes should be 9bar.
The minimum cover of the main vacuum pipeline under roads should be only 1m and 1.2m. The vacuum sewer mains and branch connections should have isolation valves ever 500m and 200m respectively.

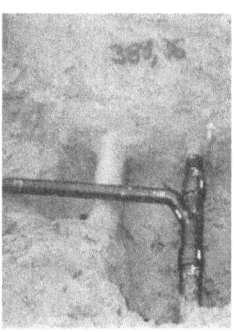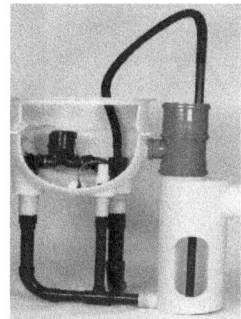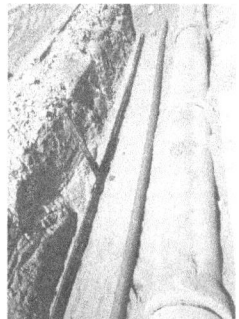

ture 11_29 to 11_31: PE-HD vacuum Pipe with individual connection (left), collection chamber (centre), wastewater collection, water supply and stormwater-drainage pipes in one trench (right)

Since the flushing velocity is provided by the suction pressure, the pipelines do not require a downward slope, although they should have a minimum gradient of 1 in 500; however, the pipes can even be laid uphill. Lifts – or short, upward sections of pipe – can be used to ensure that the pipes do not have to be laid at excessive depths or to avoid objects. It is recommended that the pipe should be laid with a saw-tooth profile.

Since the whole system is watertight, it can be installed directly in the groundwater table, in flood areas or in the same trench as water-supply lines. Unlike gravity or small-bore sewer systems, vacuum sewer-pump capacities do not have to meet the peak wet-weather flow. There is no infiltration, no exfiltration and no groundwater contamination.

Routine maintenance checks of the network are not necessary, as a change in system pressure will indicate problems. Inspection pipes, installed at distances of approximately 100m permit the insertion of inflatable balls and precise location of the problem. The vacuum station should be inspected every week, collection chambers and vacuum vessel every year and the valve diaphragm in the collection chamber needs to be changed every five years.

Because of these technical maintenance requirements and energy-supply demands, the system is not appropriate in all locations. But it can have advantages where other systems are too costly or not feasible:
- flat topography – avoiding extensive installation excavation or lifting stations
- rock layers, running sand or a high groundwater table
- areas short of water supply or poor communities that cannot afford the amount of water necessary for operating of gravity systems at scour velocities
- ecologically sensitive locations or flooding zones
- areas with obstacles to a gravity sewer route
- installation of new fresh-water network and sewerage pipes in the same trench

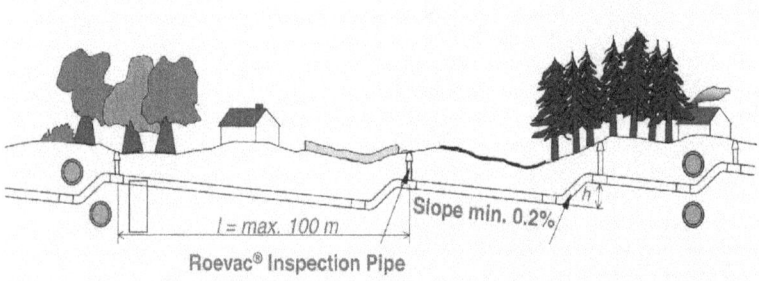

Picture 11_32: Design of vacuum sewer layout (saw-tooth profile) Credit: RoeVac® Manual, Roediger Vacuum GmbH, Germany

11.3 Sludge accumulation and treatment

All organic-degradation processes produce certain amounts of biomass or sludge, which gather at the bottom of treatment units. Sludge changes its properties with time, due to the activity of micro-organisms and the degradation of organic components. When organic degradation has been completed and all bio-chemical reactions stop, the sludge has been "stabilised". Stabilised sludge is less odorous and easier to dewater and treat.

The speed of sludge production, and the sludge's characteristics depend on the wastewater quality, sludge-retention time and other treatment parameters. Under certain conditions, there is no accumulation of sludge. A state of equilibrium between sludge production and degradation is possible in an anaerobic environment with high temperatures, adequate microbial feed within the sludge and long sludge-retention times. Under such conditions, 80% of the organic matter is converted into biogas, while the remaining organic matter is pushed out in dissolved form as effluent. Experience at existing DEWATS facilities in tropical regions, like Indonesia, shows that well-designed and constructed anaerobic units avoid the necessity of sludge removal.

In most locations, however, such boundary conditions cannot be guaranteed and even well-designed DEWATS will accumulate sludge with time. This can be caused by cooler temperatures (particularly below 15°C) or wastewater with higher mineral content. Under aerobic processes, due to the higher yield of bacteria, about 50% of the COD is transformed into biomass. Under anaerobic conditions, only about 5% of the COD is transformed into biomass, i.e. 90% less sludge (from biomass) is produced. The transformation of COD is not rate related, but yield or energy extraction related (stoichiometry, not kinetics).

The total mass of sludge is the sum of two components: non-biodegradable material in the influent and biomass produced. Sludge originating from components in the influent that can not be degraded will not be different in an aerobic or anaerobic process.

Sludge accumulation leads to a reduction of capacity and retention time within a treatment facility, ultimately resulting in inefficient treatment and the discharge of hazardous wastewater. Neglect of regular sludge removal can lead to the sludge mineralising at the bottom of the unit, until it reaches a consistency, which makes removal impossible without an operational halt and total emptying of the facility. To ensure adequate treatment and continuous operation, therefore, sludge must be removed at appropriate intervals.[37]

Sludge from domestic and husbandry wastewater is highly contaminated by worm eggs and cysts. Practices like illegal sludge-dumping into rivers, lakes or malfunctioning treatment plants pose a risk to human health and the environment. Ensuring an infrastructure for safe removal, handling and treatment, therefore, should be an integral part of town or city planning and management. It must be considered in the planning and construction of any wastewater-treatment facility.

Sludge removal, drying, treatment, selling, reuse or disposal can either be practised directly by the operator of the wastewater-treatment facility or by a service provider.

11.3.1 Sludge removal

Sludge removal should only be practised by trained personnel, as both the sludge and the gases within the facility present dangers. Particularly in anaerobic processes, methane and H_2S are produced, creating a risk of suffocation. Ventilation must be provided and open fire should be prohibited at the facility. Sludge settles in layers. The top layers contain active micro-organisms, which provide treatment by feeding on the wastewater, while the lower layers stabilise and become inactive with time. The goal of desludging is to remove only the older, bottom sludge; 30 to 50cm of active sludge should remain to ensure continuous treatment efficiency.

[37] Sludge should remain within the facility as long as possible, since stabilised sludge is easier to hand and dewater. At the same time, the sludge storage capacity must not be exceeded to ensure continuous treatment efficiency. Sludge removal intervals depend on the wastewater, type of treatment and storage capacity of the facility. Conventional tank design requires sludge removal every half to three years; ponds must be emptied every one to twenty years.

Desludging can be done with buckets, by pumping or by hydraulic pressure.

- Bucket removal is discouraged because it is impossible to withdraw only the lower sludge layers. Handling poses health risks to the operators. If practised, the workers should wear protective clothing by over their mouth, hands and feet.
- For pumping, free-flow rotary pumps are recommended to prevent clogging. The pump head is lowered to the chamber floor to remove only the oldest sludge. The pumped effluent should be visible; when the sludge becomes too light in colour, pumping should be halted to give the sludge time to flow to the mouth of the pump. Only black, stabilised sludge should be removed.
- Hydraulic desludging is practised through installed pipes at the bottom of the chamber with a diameter of at least 100 to 150mm diameter. The ductile consistency of settled and compacted sludge requires the outlet of a 2.5m-long pipe to be 0.35 to 0.50m below the normal wastewater outlet, to overcome the hydraulic loss of 15 to 20%. Sludge flow is regulated with a gate valve, which has a free opening of the full diameter, or by flexible pipes, which are lowered to initiate desludging. When not in use, these flexible pipes should be closed and locked to protect against smell and insects, while valves handles should be removed to prevent children getting up to mischief.

11.3.2 Sludge treatment

The goals of sludge treatment are:
- stabilisation
- dewatering/dehydration and volume reduction
- wastewater treatment of leachate or liquids
- pathogen destruction
- agricultural reuse or environmentally safe and hygienic disposal

Unstabilised sludge should not be dried or treated openly anywhere near where people live because of bad odour and the nuisance from flies. The origin and properties of sludge, therefore, determine which treatment should be applied: sludge from grease traps, settlers or septic tanks – also called septage – contains relatively fresh waste and has a solid content below 1%. These substances should be transported to nearby centralised treatment facilities for further stabilisation before final sludge treatment. Organic industry sludge must be removed quite frequently. Treatment in anaerobic digesters is recommended, due to its high organic content and biogas potential.

Domestic DEWATS units produce small amounts of well-stabilised sludge with good dewatering properties. In urban areas, the sludge shaved be transported to an existing centralised treatment plant. Where this is not possible, two sustainable treatment and disposal options have been identified:
- Small-scale application: the stabilised sludge can be dried on sand-beds – either directly next to the DEWATS or at a more appropriate location – and eventually the sludge can be composted and turned into agriculturally valuable humus
- Large-scale application: the construction of a decentralised sludge-treatment facility with DEWATS components. This option is only financially viable, if there are enough DEWATS or on-site treatment plants in the area to provide sufficient amounts of sludge or septage for continuous operation. In many locations, such a concept is highly beneficial, as it addresses the existing problem of septage treatment from on-site systems in the area – a very common deficiency

11.3.2.1 Small-scale application – drying and composting

The sludge from most DEWATS units is a thick liquid of approximately 3 to 5% solid content. However, the loss of a large amount of water cannot be avoided as large amounts of water are withdrawn with it, the solid content of removed sludge is closer to 2%. These large liquid volumes are difficult and expensive to transport.

For small DEWATS, therefore, stabilised sludge can be spread directly on flower-beds as fertiliser. A thin layer of sludge dries almost immediately and the slight foul smell once a year will be acceptable in most locations.

Where larger amounts of sludge cannot be transported to a more suitable drying place, drying sand-beds can be installed directly next to the treatment facility. By locating the bed approximately 40cm below the water level of the plant, hydraulic pressure can be used to distribute the sludge in a 20cm thick-layer. The bed is made up of coarse aggregate (>50mm diameter) and covered with 10 to 15cm of coarse sand.

Figure 11_33:
Sludge-drying bed and well-stabilised, small sludge cluster.

The process comprises steps:
- dewatering – filtration of water through a sand-bed. Process efficiency is a function of the filter area and depth, filter material, sludge loading and sludge properties. Total Solids (TS) can be raised from 1-5% to 15-25%
- drying wind and sun assist in natural evaporation of moisture. Process efficiency is a function of sun and wind intensity, humidity, air temperature, precipitation, sludge properties and loading depth. TS can be raised to 80%

The bottom of the drying bed should be sealed, to prevent groundwater contamination, and a slight slope should lead to drainage pipes for dewatering. In hot and dry climates, a bed can be loaded perhaps five times per year. In the case of moderate temperatures, frequent rain or high humidity, special considerations – like roofs, enclosing structures or longer drying times – are required. Banana plants can be planted in the sludge bed to make the most of moisture and nutrients.

Composting is a natural, aerobic-decomposition process, in which useful micro-organisms break down organic matter and produce carbon dioxide, water, humus and heat. Properly heaped compost reaches a temperature of up to 70°C over several weeks of maturation, thereby killing pathogens, including helminths and ova. It requires no special mechanical equipment and produces a final product, humus, which has a value as fertiliser and soil conditioner.

Parameter	Description/Comment
50% moisture content	A handful of squeezed compost should feel moist and retain its form without water dripping from it. If it is too wet, dry organic material must be added. If it is too dry, it must be watered. Protection against rain or solar radiation might be advisable.
Density of 0.6 to 0.8 (promoting aeration)	Use of stiff bulking agents or bedding promotes natural aeration. Compost pile should be turned at least once to move outside material to the inside, ensuring heat treatment of all material.
C:N ratio of 30:1	Different bulking agents can be used to adjust the ratio.
pH value of 5 to 9	Verified with litmus paper; low pH can be raised with lime or other amorphous alkaline substances.

Table 37: Requirements for a successful composting process

The largest problem with economic sludge composting is the high water content of sludge. Where large amounts of dry organic matter are available, it can be mixed with dewatered sludge (of at least 25% Total Solids) to achieve the desired Total Solids (TS) and consistency.

A successful composting process requires:

Composting can be practised within permeable boxes or elongated piles called windrows.

- If boxes are used, they must have a door for loading and removal. The walls of the box must either contain openings for oxygen supply from all sides of the compost – or the compost must be turned frequently
- If windrows are used, they should not be more than 1.2 to 1.5m high with approximately double the width and a natural slope

Since composting demands a solid understanding of the process, some experts argue that it should only be applied for greater amounts of sludge and if a composting facility already exists. Where local knowledge of the process there is with farmers or within solid waste management schemes, sludge composting can be a successful approach.

Figure 11_34 to 11_36: Box composting (left), windrow composting (middle), compost (right)

Humus, the stabilised and sanitised product of composting, is an excellent soil improver, rich in nutrients and with good moisture-retention qualities. The desludging and composting process can be planned in accordance with agricultural cycles to provide the maximum benefit to farmers.

However, premature desludging should be discouraged, as longer desludging intervals produce a safer sludge. Agreements between sludge-treatment plant operators, local farmers or organic-fertiliser producers should be encouraged. Marketing of the final product demands control mechanisms to ensure a high-quality product and might require awareness-raising activities and advertisement campaigns to promote its benefits. Alternatively, sludge compost can be used to cover landfills, or as a raw material for making items such as flower-pots, drainage trays or bricks.

If composting is not possible but sludge is to be used fresh on agricultural land, then the sludge must be put into trenches which are covered by 25cm of soil, at least. It's not suitable in areas of high ground water.

11.3.2.2 Large-scale application – sludge and septage-treatment facility

Some DEWATS, particularly those treating animal husbandry or organic industrial wastewater, produce greater amounts of sludge. If there is no sludge-treatment facility nearby, the construction of one should be considered. In most cases, such a facility will offer great benefits to the greater local community – if it is designed to also treat septage from local on-site septic tanks and pit latrines. The following example introduces such a treatment facility.

1 Project components: sanitation and wastewater treatment – technical options

Municipal sludge treatment plant (IPLT) in Mojokerto, East Java, Indonesia

Mojokerto is a town in East Java with a population of approximately 150,000. As 60 to 80% of its wastewater is treated in on-site sanitation plants (septic tanks, latrines, and grease traps) there is great septage accumulation. The emptying of the septic tanks used to be carried out by a private company, which used three to four trucks to transport and dispose of the untreated septage into a river.

In March 2005, the Municipality of Mojokerto partnered with BEST Surabaya and BORDA to initiate a septage-management and recycling project. BORDA and BEST planned the septage-disposal service and treatment facility (IPLT). The municipality is responsible for construction, while operation will be carried out by BEST in the first year and then handed over to the municipality.

Figure 11_37 and 11_38: Septage-disposal truck, polluted river – misused as dumpsite

To ensure that the collected septage actually ends up in the treatment plant, the municipality will be using an innovative financing model. The municipality sells "chips" to the community. When a septic tank is emptied, a chip is passed to the driver of the collection truck, who takes it to the treatment facility. The treatment plant is later paid according to the amount of chips it returns to the municipality.

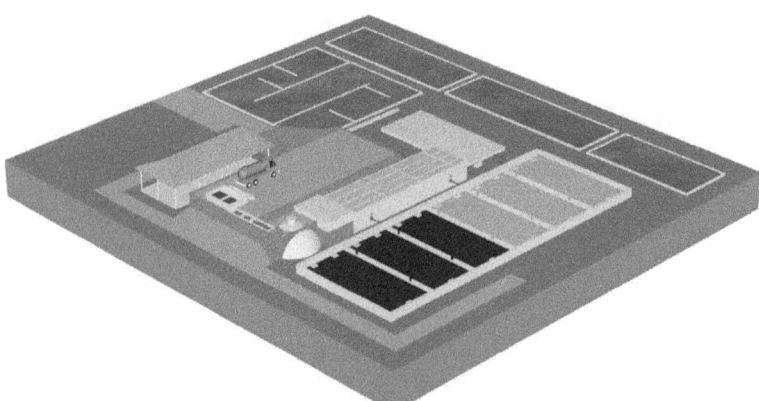

Picture 11_39 to 11_ IPLT under construction

1 Project components: sanitation and wastewater treatment – technical options

The facility is designed to handle 32m³ of septage per day. Its modules are similar to those of DEWATS and its closed components prevent odour pollution. Pre-treatment ensures that the plant is low maintenances. It includes:
- 20 and 10mm screens to avoid blockage
- a grease trap to prevent fatty accumulating sludge in pipes and reactors
- grit chambers to avoid sand accumulating in channels, pipes and reactors

A stabilisation reactor combines liquid/solid separation with anaerobic treatment. It reduces odour, oTS, COD and BOD, while improving dewaterability and drying. Biogas is produced, collected, and used/burned. The reactor consists of three chambers:
- chamber 1: mixed reactor with siphon feed to mix the sludge and promote biological activity; theoretical hydraulic retention time 1-3 days
- chamber 2: up-flow sludge bed
- chamber 3: sludge sedimentation

Design for biogas utilisation

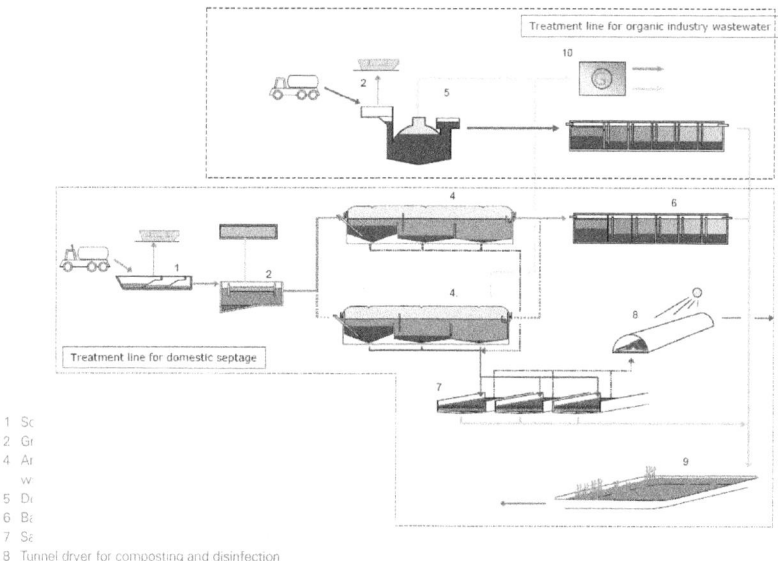

1 Sc
2 Gr
4 Ar
  w
5 Dr
6 Ba
7 Sa
8 Tunnel dryer for composting and disinfection
9 Aerobic maturation for the liquid phase
10 Biogas utilisation for producing heat or electricity

Figure 11_42:
Treatment scheme of IPLT, Mojokerto

The installation of two reactors enables alternate operation with a defined retention time for the charged sludge. The suggested sludge retention time is 15 to 20 days. A stripping column oxidises NH_4 and removes NH_3 to prevent inhibition of the anaerobic liquid treatment.

Six sand-filter beds dewater and dry the stabilised sludge. Dewatering performance ranges from 40 to 55% TS during the rainy season and from 50 to 70% TS during the dry season. Separated sludge water is drained. The sludge is composted in tunnel dryers – these consist of a simple floor with a removable greenhouse roof for storm protection. The windrows are aerated with timber air channels; the sludge is composted for 30 to 50 days. The sludge water or the liquid fraction of the sludge is treated in a baffled reactor and horizontal gravel filter.

Picture 11_43 and 11
Timber aeration channels; leachate treatment in a horizontal gravel filter

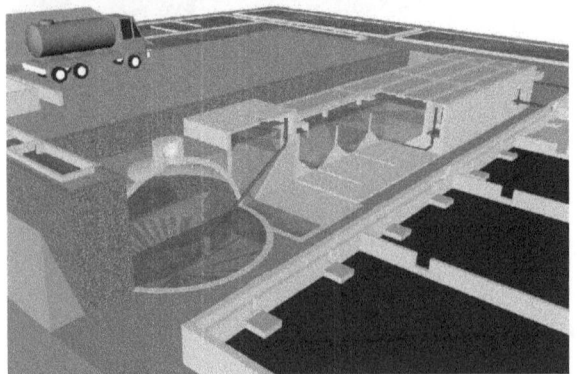

Picture 11_45:
The technical concept of IPLT Mojokerto

11 Project components: sanitation and wastewater treatment – technical options

11.4. Reuse of wastewater and sludge

11.4.1 Risks

Wastewater is never hygienically safe. Proper handling of wastewater and sludge is the only successful preventive health method. The farmer who uses wastewater for irrigation must consider the risk to his own health and to the health of those who consume the crops grown by him. He must therefore check whether the wastewater he uses for irrigation is suitable to the crops or pasture ground he intends to water.

Fresh, untreated domestic and agricultural wastewater contains over one million bacteria per millilitre, thousands of which are pathogens - both bacteria and viruses. Eggs of worms are found in the range of 1000 per litre. Epidemical statistics reveal that helminthic (intestinal worm's) infection presents the most common risk from irrigation with untreated wastewater. The risk of bacterial infection comes followed by the risk of virus infection, which is the lowest. Although the removal rates in anaerobic systems are usually over 95%, many pathogens remain even after treatment. The effluent from oxidation ponds is less pathogenic.

pathogenes	in sludge and water		in soil	on plant
	10-15° C < days	< days	20-30° C [1] < days	< days
virus	100	20	20	15
bacteria				
salmonella	100	30	20	15
cholera	30	5	10	2
fecal coli	150	50	20	15
protozoe				
amoebae cyst	30	15	10	2
worms				
ascari ova	700	360	180	30
tape worm ova	360	180	180	30

Table 38: Survival of pathogens
Source: EAWAG

[1] not exposed to direct sun light

The World Health Organisation (WHO) recommends that treated wastewater for unrestricted irrigation should contain less than 10,000 fecal coliforms per litre (1000/100ml), and less than 1 helminth egg per litre. This limit should be observed strictly since the risk of transmitting parasites is relatively high.

Pathogenic bacteria and viruses are not greatly effected in anaerobic filters or septic tanks because they remain in the treatment plant for only a few hours before they are expelled together with the liquid that exits the plant. Post treatment in a shallow pond that ensures exposure to the sun reduces the number of bacteria considerably.

Those farmers who use sewage water for farming or sludge as a fertiliser are exposed to certain permanent health risks. These health risks are controlled within organised and specialised wastewater farming or within commercial horticulture, because of certain protective measures that are taken, such as the use of boots and gloves by the workers and the transportation of the wastewater in piped systems. However, such precautions are very unlikely in small-scale farming. Plants are either watered individually with the help of buckets or trench irrigation is used. The flow of water is usually controlled by small dykes which are put together by bare hand or bare foot making direct contact with pathogens unavoidable.

A shallow storage pond to keep water standing for a day or more before it is used may minimise the number of pathogens, but would hardly reduce the indirect health risk. It is also likely that children will play here, ducks will come to swim and animals may start to drink. Fencing may help. A more foolproof preventive measure may be an establish health-education programme that reminds users of the dangers – and the precaution they need to take.

Consumers of crops grown by such means and animals that graze on pastures that are irrigated with wastewater are also endangered. Since bacteria and viruses are killed by a few hours, or at most a few days of exposure to air, wastewater should not be spread on plants which are eaten raw (e.g. lettuce) for at least two weeks prior to harvesting. India has prohibited the use of wastewater irrigation for crops that are likely to be consumed uncooked.

Since bacteria and viruses stay alive much longer when wastewater percolates into the ground, root crops like potatoes or carrots except for seeds or seedlings should not be irrigated with wastewater.

1 Project components: sanitation and wastewater treatment – technical options

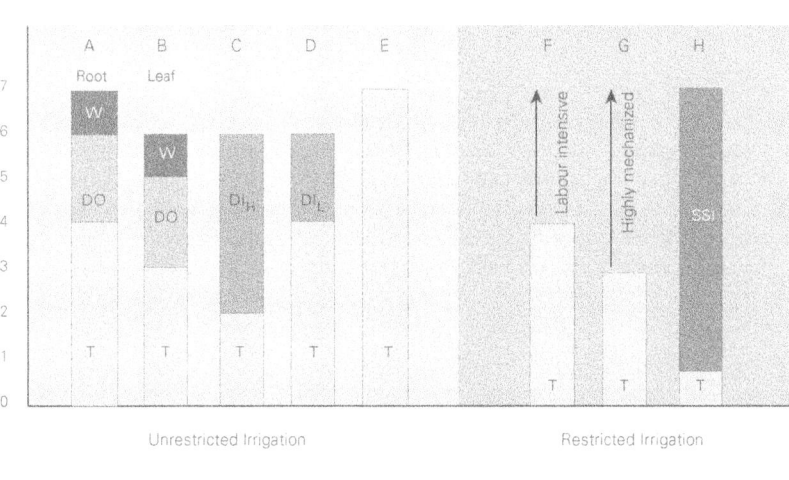

T = Treatment
DO = Die-off
W = Washing of produce
DI = Drip irrigation (H = High crops, L = Low Crops)
SSI = Subsurface irrigation

Figure 11_46: Examples of options for the reduction of viral, bacterial and protozoan pathogens by different combinations of health protection measures, Source: WHO guidelines, 2006 (see page 323)

Type of irrigation	Option (Picture 11_46)	Required pathogen reduction by treatment (log units)	Verification monitoring level (E.coli per 100ml)	Notes
Unrestricted	A	4	$\leq 10^3$	Root crops
	B	3	$\leq 10^4$	Leaf crops
	C	2	$\leq 10^5$	Drip irrigation of high-growing crops
	D	4	$\leq 10^3$	Drip irrigation of low-growing crops
	E	6 or 7	$\leq 10^1$ or $\leq 10^0$	Verification level depends on the requirements of the local regulatory agency
Restricted	F	4	$\leq 10^4$	Labour-intensive agriculture (protective of adults and children under 15)
	G	3	$\leq 10^5$	Highly mechanized agriculture
	H	0.5	$\leq 10^9$	Pathogen removal in a septic tank

Table 39: Verification monitoring of wastewater treatment for the various levels of wastewater treatment Options A–G in Picture 11_46, Source: WHO guidelines, 2006

* For example, for secondary treatment, filtration and disinfection:
five-day biochemical oxygen demand, <10 mg/l;
turbidity, <2 nephelometric turbidity units; chlorine residual, 1 mg/l;
pH, 6-9; and faecal coliforms, not detectable in 100 ml.

11.4.2. Groundwater recharge

Recharge of groundwater is probably the best way to reuse wastewater particularly since the groundwater table tends to lower almost everywhere. Wastewater had been freshwater, and freshwater drawn from wells has been groundwater before. Sustainable development is directly related to the availability of water from the ground. Thus, recharging of this source becomes absolutely vital to human civilisation. The main question is how far the wastewater needs treatment before it may be discharged to the ground. Due to the high risk of groundwater pollution, this topic is very delicate and needs to be handeld with highest precaution.

11.4.3 Fishponds

Wastewater is full of nutrients which, when directly used by algae, water plants and lower animals could become fish feed. But fish need also oxygen to breathe, which must be dissolved in water in the pure form of O_2 (4mg/l for carp species, > 6mg/l for trout species). Because free oxygen is needed for degradation of the organic matter present in wastewater, it cannot be expected to be in sufficient supply for the survival of fish. Therefore, pre-treated wastewater must be mixed with freshwater from rivers or lakes, otherwise wastewater ponds must become so large that oxygen supply via pond surface overrules the oxygen demand of the organic load.

The organic load on fishponds should be below 5g $BOD/m^2 \times d$ before 5 times dilution with freshwater. This implies that if the chances of dilution are non-existent, the organic load may be 1g $BOD/m^2 \times d$.
If possible, there should be several inlet points in order to distribute organic matter more equally where it comes into contact with oxygen quickly. As mentioned before a turbulent surface increases the oxygen intake, and cooler temperatures increase waters ability to store free oxygen. However, it is not worthwhile trying to increase oxygen intake by specialy shaped inlet structures or similar measures. At a stage where oxygen deficiency can only be little, oxygen absorption is also little.

The pH should be 7 to 8. Fish culture is not possible if wastewater may be toxic or polluted by mineral oils, temporarily or permanently.

Wastewater should not be mixed with freshwater before the fishpond. Otherwise wastewater nutrients would initiate the heavy growth of fungi, algae and other species without being consumed by fish. When a fishpond is started, it should be filled with fresh water, wastewater is added later.

When using natural lakes for wastewater-based fishery it should be known, whether the lake is legally considered to be part of the treatment system or already part of the environment in which wastewater is discharged. In other words, it must be clear whether discharge standards must be observed at the inlet or whether the effluent of the lake will do.

The type and condition of fish are an indicator of water quality. Carp can live in water with a lower oxygen content and are the most common species in wastewater-based fish culture. Tilapia has become the most common "Development Project Fish" and is also growing well in wastewater ponds. Tench species often have difficulty surviving, because they take feed from the ground and run into get problems with anaerobic bottom sludge. It is advisable to empty the ponds once a year to remove sludge or at least expose the bottom sludge to oxygen for stabilisation.

Fishponds are normally more turbid than other ponds, because fish swirl up sludge from the ground. Trout species survive surprisingly well, despite higher turbidity, when the oxygen content is sufficient. However, it should be clear that more specialised knowledge of fish species, fish production and marketing is needed than can be contained in this chapter. More information is available from the regional offices of fishery departments and should be obtained before starting a wastewater fish-farming system.

Fishponds have a hydraulic retention time of 3 to 10 days and a depth of 0.5 to 0.8m. Net fish production is in the range of 500kg/ha (50g /m^2), 900 to 1,200kg/ha are said to be harvested from Calcutta's municipality fish farm. There is also the possibility of raising fish in 2.5 to 3m deep ponds where different kinds of fish live in different strata. An almost unbelievable 12,000kg/ha are claimed to have been harvested in Brazil in such ponds every year. A higher fish population produces more sludge which reduces the amount of free oxygen. Whether wastewater-based fishery becomes a viable business depends on the market price of fish and fishery operating costs. Fingerlings must be kept separate because fish, when set free, should weight 350g in order to be too heavy for fishing birds.

Losses can reach 50% when fishponds become an ecological niche which attracts fish-hunting birds.

Fish lose the foul taste of wastewater if they are kept for a few days in fresh water before consumption. This also reduces the risk of pathogen transfer. Fishermen need to be aware that the wastewater always bears a certain, albeit small, health risk.

11.4.4 Irrigation

Treated domestic or mixed community wastewater is ideal for irrigating parks and flower gardens. Irrigation normally takes place in the evening or early morning so that people won't be bothered by the slightly foul smell of anaerobic effluent. Nonetheless, the irrigation of public parks is often forbidden by law.

In order to provide updated and reliable orientation, the World Health Organisation has published in 2006 four volumes of "Guidelines for the safe use of wastewater, excreta and greywater":
Volume 1 - Policy and regulatory aspects
Volume 2 - Wastewater use in agriculture
Volume 3 - Wastewater and excreta use in aquaculture
Volume 4 - Excreta and greywater us in agriculture
These documents are widely recognised and can be downloaded from the WHO website free of charge.

A few facts to bear in mind:
For an irrigation rate of 2m per year (20,000m^3/ha) – the normal requirement in semi-arid areas – even welltreated wastewater with concentrations as low 15mg/l of total nitrogen and 3mg/l total phosphorus provides 300kg N and 60kg P per ha via irrigation without additional cost; at the same time the respective amount of groundwater is saved.

In areas where there is plenty of rain, less water is needed for irrigation. So pre-settled but otherwise fresh wastewater may be more appropriate with respect to fertiliser. With 0.1m per year (1,000m^3/ha) of fresh wastewater for irrigation, some 60kg nitrogen, 15kg phosphorus and a similar amount of potassium could be applied per ha. However, domestic wastewater in modern households some-

1 Project components: sanitation and wastewater treatment – technical options

times lacks the potassium which might need to be added to mobilise nitrogen and phosphorus.

As this book deals with wastewater, it does not provide detailed information on either general or specific local questions of agriculture or the nutrient requirements of different crops. Each farmer has to find out his or her own preferred method and his or her own way of using efficient and safe quantities of water. The practical farmer knows which nutrients are needed for which crop and a trained agriculturist would also know – from wastewater analysis – whether the composition of nutrients and trace elements suits the proposed plantating. He or she will also know, from that analysis, whether too many of toxic elements remain in the water (toxic elements might play a role if the COD is much higher than the BOD). Such tests are advisable when using treated industrial or hospital wastewater for the first time. The person responsible for the wastewater source is obliged to inform farmers about toxic or otherwise dangerous substances in the effluent, for example, radioactive elements from x-ray laboratories.

Original saline water will remain saline even after intensive treatment. Copper and other metals, especially heavy metals, accumulate in the soil. Long-term application of such water will spoil the soil forever.

11.4.5 Reuse for process and domestic purposes

Pathogenic wastewater – from domestic sources, slaughterhouses or animal stables should not be reused for any purpose, except irrigation. Partly treated organic wastewater (this is more or less all wastewater from DEWATS treatment) should not be reused directly as process water in industries or as flushing water in toilets. Reusing wastewater will always mean that some traces of organic matter or toxic substances remain or accumulate. Reuse also means longer retention times in a closed system which might facilitate anaerobic processes within pipes and tanks which will cause corrosion. There is also a theoretical risk of biogas explosion.

To suppress organic decay one may have to add lime, which might form limestone inside the system or other inhibiting substances which would make appropriate final wastewater treatment costly. For example, even the first washing water in a fruit-processing plant or a potato-chip plant might already contain too much organic matter for any reuse without lime being added to suppress fermentation.

The chance of re-circulating parts of the water to serve the production process is limited, especially when the wastewater engineer and the production engineer don't have the necessary knowledge. The pollution content and level of treatment needed, as well as the amount of water required for consumption – and the wastewater flow over a given period of one day (or one season) – must be investigated. It might be necessary to build intermediate water stores and install additional pumps. Wastewater reuse is an option that deserves close consideration in the context of sustainable development. But, for process and domestic purposes, the accompanying problems mean it can't be recommended.

Reusing industrial wastewater which is only slightly polluted and perhaps not organically polluted, is a completely different matter. For example, press water in a soap factory may be reused for mixing the next load of soap paste. All water-consuming modern industries have reduced their water consumption considerably in the last few years. In most countries, including India and China, water-consumption limits are obligatory for many industrial processes, such as sugar refining, brewing, canning, etc.: Saving water in the process is always better than reusing water which has been carelessly wasted and polluted.

11.5 Biogas utilisation

11.5.1 Biogas

All anaerobic systems produce biogas. 55 to 75% of methane (CH_4), 25 to 45% of carbohydrate (CO_2) plus traces of H_2S, H, NH_3 go to form biogas. The mild but typical foul smell of biogas is due to the hydro-sulphur, and after which transforming into H_2SO_3, is also responsible for the corrosive nature of biogas. The composition rate of biogas depends on the properties of wastewater and on the design of the reactor – the retention time. Theoretically, the rate of methane production is 350l per kg removed BOD_{total}. In practice however, methane production should be compared to 1kg removed COD of where values are closer to the removed BOD_{total} than to the removed BOD_5. By doing so, one assumes that during anaerobic digestion only biodegradable COD is removed, which is involved in the production of methane. In reality, the gas production rates are lower than this because a part of the biogas dissolves in water and cannot be collected in gaseous form. It is also the norm to relate biogas production to organic dry matter (DM) in case of very strong viscous substrate, 300 to 450l biogas per kg DM can be expected.

1 Project components: sanitation and wastewater treatment – technical options

The calorific value of methane is $35.8 MJ/m^3$ ($9.94 kwh/m^3$). The calorific value of biogas depends on the methane content. Hydrogen has practically no role. As a rule of thumb, $1m^3$ biogas can substitute 5kg of firewood or 0.6l of diesel fuel.

industry	COD per product kg/to	COD removal %	relative gas production $m^3 CH_4/COD_{in}$	methane content %
beet sugar	6-8	70-90	0.24-0.32	65-85
starch – potato	30-40	75-85	0.26-0.30	75-85
starch – wheat	100-120	80-95	0.28-0.33	55-65
starch – maize	8-17	80-90	0.28-0.32	65-75
molasses	180-250	60-75	0.21-0.26	60-70
distillery – potato	50-70	55-65	0.19-0.23	65-70
distillery – corn	180-200	55-65	0.19-0.23	65-70
pectine		75-80	0.26-0.28	50-60
potato processing	15-25	70-90	0.24-0.32	70-80
sour pickles	15-20	80-90	0.28-0.28	70-75
fruit juice	2-6	70-85	0.24-0.30	70-80
milk processing	1-6	70-80	0.24-0.28	65-75
breweries	5-10	70-85	0.24-0.33	75-85
animal slaughter	5-10	75-90	0.26-0.32	80-85
cellulose	110-125	75-90	0.26-0.33	70-75
paper/board	4-30	60-80	0.21-0.28	70-80

e 40:
otential biogas
roduction from
ome selected in-
ustrial processes.
ource: ATV, BDE,
KS

11.5.2 Scope of use

As pointed out above, biogas may be used in burners for cooking or in combustion engines to generate power. Its use will depend on whether enough can be supplied regularly to meet the minimum requirement of a particular use. If biogas cannot be utilised, it should be released in the air via safe ventilation or flaring. It is pointless to collect, store and distribute biogas when there is no real demand.

It is not essential to extract carbohydrate (CO_2) before biogas is used. But it might be advisable to remove an unusually high H_2S content with the help of iron oxide: Biogas flows through a drum or pipe filled with iron oxide (e.g. rusted iron borings or swarf). The oxygen reacts with the hydrogen to form water, while sulphur and iron (or sulphide of iron) remain. The iron may be reused if it becomes rusty again from exposure to air.

The minimum amount of biogas for a household kitchen requires is approximately 2m^3/d. Approximately 20 to 30m^3 of domestic wastewater is required daily to produce the minimum amount of gas. From an economic point of view, biogas utilisation from wastewater becomes meaningful if the strength of the wastewater is at least 1,500mg/l COD and the regular daily flow is 20m^3.

The best use of biogas is for heat production. Biogas burners are simple in principle and can be made from converted LPG-burners. Biogas can be used for cooking in the home and canteens, or for drying and heating as part of industrial processes. The very best use of biogas would be as fuel for the same process that produces the wastewater.

Biogas can also be used in gas lamps. But the light from a biogas lamp cannot compete with an electric light.

Biogas can be used as fuel in diesel and petrol engines. As the ignition point of biogas is rather high, it will not explode under the pressure of a normal diesel engine. So, around 20% of diesel must be used for ignition, together with biogas. Diesel engines are the most suitable because they don't have to rely on a regular supply of biogas. Also, the slow flame speed of biogas is better suited to the slowly revolving diesel engine than to petrol engines. Biogas would not have enough time to burn completely with engines that run with more than 2,000 revolutions per minute.

11.5.3. Gas collection and storage

Biogas is produced within wastewater and sludge, from which it rises in bubbles to the surface. The gas must be collected above the surface and stored until it is ready for use. Even when gas production is regular, the accumulation of useable gas is irregular. Gas bubbles cause turbulence which leads to the explosive release of gas in a chain reaction. Stirring substrate, especially stirring sludge, has a similar effect. As a result of this, gas production fluctuates by plus/minus 25% from one day to the next. The volume of gas storage must provide for this fluctuation.

The volume of gas in stock changes according to gas production and the pattern of gas consumption. With rigid structures, the volume of the storage tank either changes as the volume of gas present changes, or the gas pressure increases along with the stored volume. In fixed-dome plants, and with flexible material such as plastic foils, both the volume and the pressure fluctuate.

There are two main systems for rigid materials:
- the floating drum and
- the fixed dome

For flexible material there are two variants, as well:
- the balloon, and
- the tent above water

The **floating drum** (see Pictures 9_7-C, page 183 and 11_47) is a tank that floats on water, the bottom of which is open. The actual storage volume changes depending on the amount of gas available and the drum rises above the water according to gas volume. The drum is normally made out of steel. To avoid corrosion, materials such as ferro-cement, high-density polyethylene (HDPE), butyl rubber, ethylene propylene diene monomer rubber (EPDM) and fibreglass have also been tried. As a rule, only very experienced workshops have been successful with these materials. Most find leakage a problem. The gas pressure is created by the weight of the drum (the weight is divided by the occupied surface area to calculate the pressure). A safety valve is not required as surplus gas is released under the rim when the drum rises beyond a certain point.

The **fixed dome** principle (see Picture 9_7-A+B, page 183) has been developed for biogas digesters for rural households as an alternative to the floating drum with its corrosion problem. The fixed-dome plant follows the principle of displacing liquid substrate through gas pressure. The gas pressure is created by the difference in liquid level between the inside and outside of the closed vessel. If there is very high gas pressure, the outlet pipe functions as a safety valve. The inner level of the outlet pipe, therefore, must be lower than that of the inlet.

Picture 11_47: Floating drum plant. The drum are being lifted re-painting wh allows a view the double-ring of the water ja Constructed by LPTP and BOR for a slaughter house in Java/ Indonesia.

In biogas plants that have a relatively high gas production compared to the volume of substrate, an expansion chamber is needed to sustain gas pressure during use. In the case of wastewater, where the volume of water is relatively large compared to the volume of gas production, an expansion chamber may not be required because the in-flowing wastewater replaces the wastewater, which has been pushed out by the gas. For this reason, gas consumption must correspond with intensive wastewater inflow. An expansion chamber is required when there is little to no wastewater flow during gas consumption. It is not when the simultaneous volume of consumed gas is less than the volume of wastewater inflow.

The surface area of an anaerobic treatment tank is relatively large compared to the amount of biogas produced. Consequently, fluctuation in liquid levels as a result of variation in gas volumes in the upper part of the reactor are relatively small. All the same, it may influence the design, especially the level of baffles needed to retain floating solids.

1 Project components: sanitation and wastewater treatment – technical options

Biogas is an end-product of decomposition and, therefore, has very fine molecules that can pass through the smallest crack and the finest hole. So its storage must be as gas-tight as a bicycle tube. The usual-quality concrete and masonry is not sufficiently gas-tight – bricks are porous and concrete has cracks. So, bricks and concrete must be well plastered by applying several layers and adding special compounds to the mortar to minimise shrinking rates. Several layers of plaster help to cover the cracks on one layer with the next layer of plastering, in the hope that the cracks in different layers do not appear at the same spot.

1st layer	cement–water brushing
2nd layer	cement plaster 1:2.5
3rd layer	cement–water brushing
4th layer	cement plaster 1:2.5 with water-proof compound
5th layer	cement–water brushing with water proof compound
6th layer	cement plaster 1:2.5 with water-proof compound
7th layer	cement–water finish with water-proof compound

Table 41:
Typical prescription for gas-tight plaster in fixed-dome biogas plants. The method was developed by CAMARTEC/GTZ in Arusha, Tanzania and has been successfully applied in many countries since 1989.
Source: Camartec, BORDA

Picture 11_48:
Fixed-dome biogas plant for cattle dung, under construction by CYSD and BORDA at a farm in Orissa, India

A structure under pressure cannot develop cracks, therefore the structure of the gas storage should be under pressure whenever possible. This is the reason why anaerobic reactors should have arched ceilings, So that a heavy-soil covering creates the required pressure. Normally, baffled reactors and anaerobic filters are rectangular. As it is difficult and expensive to make these structures gas-tight, and considering the fact that gas production is greatest in the first part of the reactor, it may be reasonable to collect gas from the first chambers only. These chambers must be completely gas-tight; rear chambers must be ventilated separately.

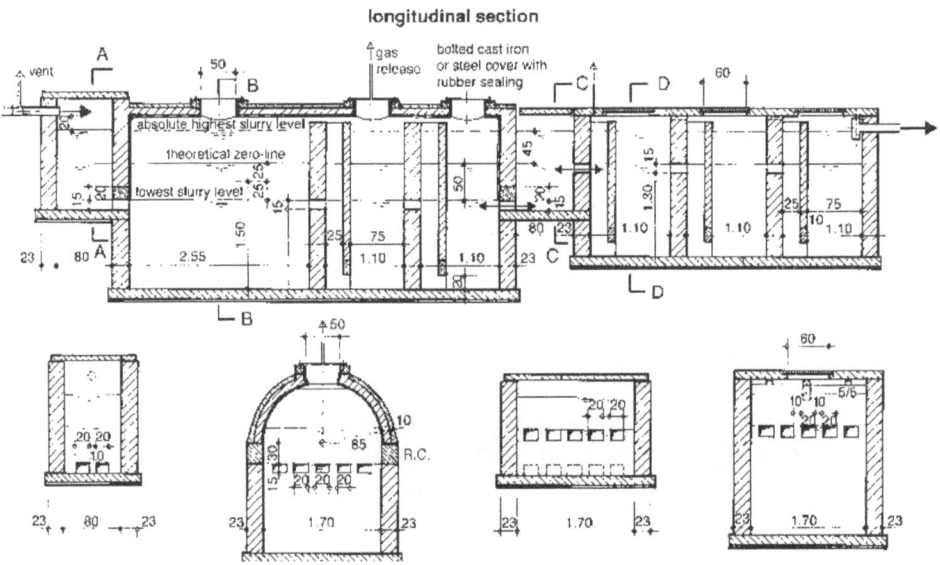

Picture 11_49:
Baffled septic tank with biogas utilisation. Only biogas from the settler and the first two baffled chambers is used. They are arched in order to guarantee a gas-tight structure. The tanks which store biogas are separated from the three chambers at the rear of which biogas is not collected. The design is based on 25m³ daily wastewater flow, 4,000mg/l COD and a necessary gas storage volume of 8m³.

Tent systems are mostly used with anaerobic ponds. Balloons may be connected to any anaerobic-tank reactor. Balloons and tent systems require the same material. These materials must be gas-tight, UV-resistant, flexible and strong. PVC is not suitable. The weakest points are the seams and in particular the connections between the foil and the pipes. To secure gas tightness, foils of tent plants are fixed to the solid structure below the liquid level. Foil covering may also be fixed to frames floating on the wastewater. Balloons should be laid on a sand bedding or hung on belts or girdles. It may be necessary to protect them against damage by rodents. The gas pressure must be kept under control to match the permissible stress of the material, especially at joints. Fitting a safety valve, which functions as a water seal on gas pressure, should solve this problem.

Balloon and tent systems, unless securely fenced and protected against stones or rubbish thrown by children, are not suitable for domestic plants.

Figure 11_50:
Tent gas storage above a liquid manure tank. Biogas plant constructed by SODEPRA and GTZ at a cattle park in Ferkessedougou, Ivory Coast.

11.5.4 Distribution of biogas

Normal water-pipe installation technology may also be used for biogas distribution in DEWATS. But ball valves should replace gate valves. All parts should be reasonably resistant against corrosion by sulphuric acid. The joints in galvanised-steel pipes should be sealed with hemp and grease or with special sealing tape. Joints of PVC pipes must be glued; the glue must be spread around the total circumference of the pipe.

Biogas always contains a certain amount of water vapour, which condenses to water when the gas cools down. This water must be drained; otherwise it may block the gas flow. Drain valves or automatic water traps to avoid blockages must be provided at the lowest point of each pipe section. Pipes must be laid in a continuous slope towards the drain points; straight horizontal pipes should not sag.

Gas pressure drops as the pipe gets longer, and more so with smaller pipe diameters. The pipe diameter must be larger when the point of consumption is far off. Long distances are generally not a problem, but should be kept as short as possible for economic reasons. Connecting stoves or lamps with a piece of flexible hose to the main distribution pipe means that equipment can be moved without disconnecting the pipe. It also allows for condensed water to be drained. In the case of fixed-dome plants, where it is difficult to see the amount of gas available, a U-shaped gas-pressure meter (manometer) could be installed near the point of consumption.

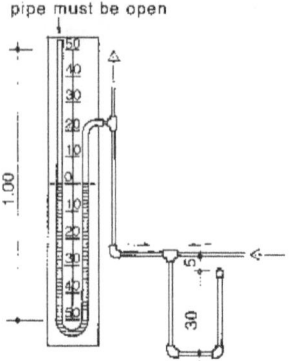

Picture 11_51: Pressure gauge out of transparen flexible pipes and water trap to coll condensed vapou which develops in gas pipes due to changing temperature. Water must be drawn from the trap whe gas-flames start t flicker.

11.5.5. Gas appliances

In principle, biogas can be used in the same way as any other gaseous fuel, for example in refrigerators, incubators, or water heaters. But it's most commonly used in stoves, lamps and diesel engines.

Biogas needs a certain amount of air to burn – on average, one cubic metre of gas requires 5.7m³ of air for complete combustion, a quarter of what LPG would need. So, LPG burners have smaller jets; consequently the relative air intake compared to biogas burners is greater. The air intake needed for combustion is regulated by the difference of jet diameter to mixing-pipe diameter. For open burners, which draw primary air at the jet and some secondary air at the flame port, the ratio between jet diameter and mixing-pipe diameter may be taken as 1:6. For lamps, where secondary air supply is lower, this figure may be 1:8.

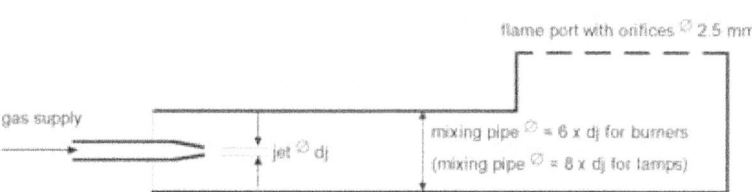

ture 11_52: Principle design parameters for biogas appliances. The relation between jet diameter and mixing-pipe diameter is important for good performance and efficiency, irrespective of gas pressure. Other parameters are less crucial or can be found by trial and error, for example, the number and diameter of orifices or the length of mixing pipe.

When converting LPG equipment to biogas, the jet must be widened, to around one-sixth of the diameter of the mixing pipe of a burner. These ratios are the same for all gas pressures. There is no need to regulate the air intake when gas pressure changes. However, air requirement is greater when methane content is higher. The difference is too small to be of practical importance. Since the flame speed of biogas is relatively low, biogas flames tend to be blown off when gas pressure is high. It may be advisable to increase the number or size of orifices at the flame port in order to reduce the speed. It is also possible to reduce the flow by placing an obstacle at the flame outlet; for example, a pot set on the burner.

It is trickier to regulate the air-gas mixture in lamps that use textile mantles, because the hottest part of the flame must be directed at the mantle so that the mineral particles glow. If the flame burns inside the mantle, the pressure might be too low and the primary air may be too much. If, on the other hand, the flame burns outside the mantle, there would not be enough primary air and the pressure might be too high. As the composition of biogas also has a role to play, it is not easy to give general recommendations for lamp design. Practical testing is the only solution.

Diesel engines always have a surplus of air and proper mixing is not required. The gas is connected to the air – supply pipe after the air filter. The mixing of air and gas is improved when gas enters the air pipe by cross flow. Dual fuel engines are started with 100% diesel; biogas is added slowly when the engine is hot and under load. The amount of biogas is regulated by hand. The engine usually starts to splutter when there is too much gas. When the engine runs smoothly, it is regulated like a pure diesel engine with the help of the throttle. For generating 1kwh electricity, approximately 1.5m^3 biogas and 0.14l diesel are required.

2 System malfunction – symptoms, problems, solutions

DEWATS are designed to be particularly robust. Nonetheless, problems may be caused by improper use or operation, insufficient maintenance or structural flaws. A malfunctioning system is a risk to public health and the environment. Reoccurring problems create further complications, if they are not quickly attended to.

As a result, each DEWATS facility requires responsible personnel to:
- recognise the symptoms of a malfunctioning system at an early stage
- identify the cause of the problem
- repair the system, appropriate measures, as soon as possible

There are two main types of system malfunction:
- insufficient treatment of wastewater and
- reduced flow at the outlet of the facility

In case of malfunction, the following sections can be consulted for guidance. They present common symptoms and list possible problems and specific maintenance solutions.

To facilitate troubleshooting, it is beneficial to have a plan of the system and a record of past maintenance activities. Records of pumping, inspection, and other maintenance work should be kept (see operation & maintenance manual referred to in picture 6_25, page 128).
It should be clear who is responsible and who can be contacted if the problem reoccurs. A list of specialists (including name, address and phone numbers) should be available – and all staff and users know where it is kept.

12.1 Insufficient treatment of wastewater

Treatment of the wastewater is considered insufficient if it does not correspond to the desired discharge standards in one or several of the following categories:
- BOD
- COD
- suspended solids
- smell
- faecal contamination

Symptoms
• extensive plant growth (eutrophication) in the discharge water body • fish dying • turbid effluent • frothy discharge • biological and nutrient contamination in nearby wells or surface waters • smell • high pH-value

Problem	Solution
Accumulated sludge within Imhoff tank, septic tank, anaerobic baffled reactor, biogas digester or pond system. This leads to a reduction of the hydraulic retention time for treatment.	determining sludge depth: 1. wrap one metre of white fabric around the end of a long stick 2. place the stick into the sludge, behind the outlet baffle – leaving it there for one minute 3. remove the stick and note the sludge line 4. If the sludge line is within 30cm of the outlet baffle or 45cm within the outlet fitting, the system requires cleaning. For details on correct sludge removal, handling, treatment and reuse – see section 11.3. If many non-biodegradable materials such as plastics, disposable nappies or sanitary towels are found in the sludge – awareness raising for proper use of the system is necessary.

Table 42: Insufficient treatment of wastewa

2 System malfunction – symptoms, problems, solutions

Table 42: Insufficient treatment of wastewater

Problem	Solution
Accumulated scum (soap suds and fat) floating on the wastewater surface within Imhoff tank, septic tank or baffled reactor Scum reaches the tank outlet and flows into subsequent treatment units, which are not designed to handle it. Extremely fast scum growth may be an indicator of excessive hydraulic loading.	Measuring scum depth: 1. attach a 15cm square board to the bottom of a stick. 2. extend the stick through the scum to locate the bottom of the baffle or effluent pipe in the tank – mark the stick to indicate that point. 3. raise the stick to locate (by feeling or seeing) the bottom of the scum layer – mark the stick again. 4. If the marks are less than 8cm apart, or if the scum surface is less than 3cm from the top of the outlet baffle, the tank requires cleaning. Treatment facilities treating very greasy wastewater (for example from restaurants) should have a grease trap (see section 9.2, page 176) – remove grease from the water surface with a shovel and deposit it in grease pit (at least 10m from well) twice weekly.

cont. Table 42: insufficient treatment of wastew[ater]

Problem	Solution
Excessive inflow quantity caused by • increased number of users • changed user habits • structural deficiencies This leads to a reduction of the hydraulic retention time; insufficient time for treatment can lead to low pH levels, caused by volatile fatty acids. It can also lead to backlogging water within the system, or extrusion of water at unforeseen places, if the filter velocity though wetland or filters is insufficient.	Uncontrolled inflow of ground- or stormwater through leaking or damaged pipes or structures must be prevented by locating infiltration points and carrying out repairs. (This can include leaking roofs of community sanitation centre shower or toilet rooms). Uncontrolled stormwater inflow through maintenance openings must be prevented. Attaching wastewater flow from more users than the system was designed for must be discouraged. If the wastewater amount has grown beyond system capacity, a system upgrade is necessary or a parallel treatment system must be installed. Alternatively, awareness-raising activities to promote water-saving habits or fixtures can be applied.
Daily peaks higher than expected	Consider an equalisation tank.

2 System malfunction – symptoms, problems, solutions

Table 42: insufficient treatment of wastewater

Problem	Solution
Excessive inflow contamination caused by:	Inflow of inappropriate wastewaters must be prevented.
• Inflow of wastewater sources unforeseen in the planning of the facility (e.g. industrial wastewater connected to a domestic wastewater treatment unit).	An appropriate facility or an upgrade of the existing treatment plant is required.
• Excessive BOD and ammonia loadings.	**Anaerobic ponds:** adding lime ($12g/m^3$ of the pond) may help to raise the pH value.
Can lead to increased accumulation of settleable solids, low pH-value due to volatile fatty acids or temperature shifts in anaerobic reactors (esp. in the case of illegal industrial connection) Methanogenesis is sensitive to both high and low pHs and occurs between pH 6.5 and pH 8. Low pH-levels are inhibiting methanogenic organisms and causing smells.	**Facultative ponds:** create multiple inlets to the pond. Periodically add sodium nitrogen as a supplement source of combined oxygen. Where possible, public-awareness campaigns can help to minimise pollution through habit change, for example in cooking practices or handling of kitchen waste.
System short circuit caused by • defective separation walls and baffles in tanks or reactors • excessive aquatic vegetation in facultative ponds, reducing the area of flow across the system This leads to less retained settleable solids and reduction of the hydraulic retention time (also see "incorrect retention time" below).	In most cases, draining the facility is necessary to carry out the required repairs or maintenance.

cont. Table 42: insufficient treatment of wastewa

Problem	Solution
Incorrect retention time within the unit can create smell or effluent-quality problems In grit chambers, a rotten-egg smell indicates sedimentation of organic matter, due to slow flow velocity/too-long retention time. The removed sand is grey and contains grease. In anaerobic ponds, HRT longer than one day leads to fermentation – not only of the bottom sludge but also the liquid phase. A too-short HRT creates effluent with low pH and emits H_2S odour. In facultative ponds, growth of filamentous algae and moss indicates underloading. Poor flow distribution can be responsible for insufficient retention time and treatment.	Adjustments of flow must be made: • Increasing flow velocity by using fewer parallel units, if available. • Lowering retention time by bypassing overloaded units if the following ones can handle the higher load. Ideally, upgrading of the facility. • Increasing retention time by redu-cing flow quantity or capping peakflow (equalisation tank) Check inlets and distribution of flow to treatment units like ponds or wetlands: • **Anaerobic ponds:** distribution by perforated pipes on the bottom of the pond. • **Facultative ponds:** create several inlets with uniform distribution to each. • **Wetlands:** ensure influent distribution across the full width.
Incorrect water level in horizontal gravel filters, resulting in surface algae growth or insufficient treatment.	The water level should be just below the filter surface; the flow-regulation pipe should be adjusted accordingly, during weekly maintenance tasks.

2 System malfunction – symptoms, problems, solutions

Table 42: Insufficient treatment of wastewater

Problem	Solution
Scum layers or floating material on ponds can hinder some treatment processes.	**Anaerobic ponds:** no measure needs to be taken. The scum layer helps to maintain the absence of oxygen, controls the temperature and prevents the release of bad odours. **Facultative ponds:** remove scum layers, place scum into plastic bags and practise proper garbage disposal. Light and wind penetration of the pond surface should be ensured.
Growth of aquatic or terrestrial vegetation or algae in or on ponds can hinder the treatment process and create smell.	**Anaerobic ponds:** Vegetation on internal or external slopes, as well as in shallow water should be removed completely and regularly. **Facultative ponds:** remove excessive algae growth on the surface, which is prohibiting passage of light, with sieves. Remove excessive aquatic plants restricting the area flow and creating oxygen demand upon plant mortality. **Indicator ponds (polishing ponds):** Algae should be removed from the walls by a brush every 14 days.
High concentrations of algae (SS) in the effluent of pond systems	Install baffles to retain and remove algae. Use multiple cells in series with shorter retention time in each.

cont. Table 42: insufficient treatment of wastewa[ter]

Problem	Solution
Cloudy weather and low temperature over long stretches of time reducing treatment efficiency in facultative ponds and causing bad odours.	Reduce the depth of the facultative pond temporarily. If possible, put ponds in parallel operation.
Metal or concrete erosion in anaerobic reactors caused by insufficient ventilation.	Check and remove obstructions to the ventilation system, including chamber connections.
Insufficient water seal in the biogas settler causing inefficient treatment and making the system unsafe.	Insert a stick through the hole in the manhole cover to measure the distance until it gets wet. If necessary, refill water seal.

If a system malfunction was caused by improper use of the system, awareness raising campaigns should teach users how to prevent such problems in the future.

If a system malfunction was caused by insufficient operation and maintenance the existing maintenance schedule should be reviewed and adhered to in the future. A maintenance time schedule and log book is recommended (see operation & maintenance manual, picture 6_25, page 128).

12.2 Reduced flow at the outlet of the facility

The effluent volume of a system does not always equal the influent volume – it depends on the amount of evaporation of constructed wetlands or pond systems. However, when the amount of effluent is far less than expected, the system is either clogged at one or more locations and/or is discharging wastewater at unforeseen locations. All control openings should be checked to identify the location causing the irregularity in flow.

Symptoms
- poorly draining toilets, showers, sinks or drains – nuisance to the users, easily identifiable
- extrusion of wastewater at unforeseen places – environmental & health hazard, likely to go unnoticed or to be disregarded. Noticeable as:
  - pools of water in unexpected places
  - lush, green vegetation, even during dry weather, in places where there should be none
  - pathogen or nitrate contamination of nearby wells
  - dying plants in a horizontal gravel filter, due to lack of water
- reduced flow at the outlet or significant fluctuations – parameter should be monitored by maintenance personnel

Problem	Solution
Pump malfunction, hindering wastewater flow	If a pump is used, check for obstructions and remove them. Check whether the pump-level control is functioning and that the pump is adequately lubricated. Each pump differs slightly, so consult the maintenance manual for the pump for more information about pump maintenance.

cont. table 43: reduced flow at outlet of the facil

Problem	Solution
Clogged pipes – anywhere between the household and location of effluent discharge, including wastewater-treatment plant possible causes include: • improper system use as garbage-disposal for non-biodegradable materials such as plastics, disposable nappies, sanitary napkins, etc. • plant roots growing into the system	Obstructions at manholes should be removed with a shovel and bucket until normal flow is achieved. Pipes should be opened at all maintenance openings to check for backlogged water. The section of clogged pipe lies between the last control opening with backlogged water and its downstream opening. The intermittent section of piping is cleared using boiling water, a drain snake or long pole. Caustic drain openers should not be applied. The reason for pipe obstruction should be identified to prevent identical problems in the future: • roots or saturated soils in the system indicate a damaged pipe. The section of pipe should be replaced. Reasons for pipe damage should be identified. Responsible trees should be removed and/or heavy loading of the pipe with machinery or vehicles should be prevented. • future system misuse should be discouraged through awareness-raising campaigns for users.

2 System malfunction – symptoms, problems, solutions

table 43: reduced flow at the outlet of the facility

Problem	Solution
Damaged pipes – anywhere between the household and location of effluent discharge, including wastewater-treatment plant Leaking pipes cause reduced flow in the system and pollute the environment. At times of high groundwater, or during strong rainfall, inflow to the damaged pipe can lead to large fluctuations of flow. Possible causes include: • plant roots growing into the system • unforeseen heavy loading (vehicles or machinery) on laid pipes • leaking joints	Monitoring flow at various control openings helps to locate leaks. Damaged pipes must be replaced. The reasons for pipe damage should be identified to prevent identical problems in the future: • trees responsible should be removed • excessive loading of the pipe with machinery or vehicles should be prevented • ensure compacted clean sand bed under the pipes and backfilling with clean granular sand, compacted in layers • during regular maintenance, check for leaking system components
Clogged anaerobic filter Inefficient treatment – as discussed in the previous chapter – results in too many suspended solids reaching the filter.	Filter material must be washed with high hydraulic pressure. In most cases the filter material must be removed, cleaned and replaced. Personnel must wear mouth and skin protection. A clogged filter is an indicator that prior treatment is insufficient and too many suspended solids reach the unit. To prevent identical problems in the future, the cause of insufficient treatment can be identified and corrected with the help of the previous section/table.

cont. table 43:
reduced flow at the outlet of the facili[ty]

Problem	Solution
Clogged horizontal gravel filter Plant growth on only certain parts of the filter can indicate irregular flow – leading to a reduction of retention time until the filter is totally clogged. Possible causes include: • any of the reasons listed in the section "inefficient treatment" – too many suspended solids reach the filter • improper use of the system (great amounts of grease or cooking oils can solidify within the filter and cause clogging) Dead plant matter or extensive weed growth on the filter surface can as well be responsible for filter clogging.	Inlet and outlet pipes and channels should periodically be checked for obstructions and cleaned, so that a uniform flow (vital for efficient treatment) can be guaranteed. The plants growing on the filter should be trimmed regularly to not less than 1m. Dead-leaf litter in and around the planted gravel filter should be manually removed every week. The area around the filter should be weeded regularly. If plants grow too densely, they should be thinned out. Look for evidence that heavy equipment has been on the wetland, filter or drainage field, to locate areas of possible compaction and damage. Identification of the cause for clogging may require digging up a small portion of the wetland or drainage field. Maintenance might require draining of the unit, material removal and cleaning. A clogged filter is an indicator that prior treatment is insufficient and too many suspended solids are reaching the unit. To prevent identical problems in the future, the cause of insufficient treatment can be identified and corrected with the help of the previous section.

2 System malfunction – symptoms, problems, solutions

cf. table 43: reduced flow at the outlet of the facility

Problem	Solution
Structural deficiencies – cracks and leaks • cracked or improperly sealed walls or floors of treatment units • leaking pipes or pipe joints • loss of water due to flaws in the liner of a horizontal gravel filter – indicated by dying plants	Testing for leaks or cracks: • Filling the unit with closed outflow; waiting for 24 hours to see if it loses water • Empty the unit with closed inflow; waiting for 24 hours to see if water infiltrates from the outside If so, locate the leaks and repair them.

12.3 Other problems and nuisances

Symptoms	
• Excessive mosquito breeding	
Problem	**Solution**
Stagnant water turns into a breeding ground for mosquitoes, which cause discomfort for those near the pond, and increase the likelihood of insect-borne diseases such as malaria.	Increase flow, so that water does not become stagnant. Alternatively, introduce lung-breathing fish into the pond (i.e. *Gambusia spp.*).
Clogged biogas lines caused by an accumulation of condensed water.	The valve to release water vapours should be opened daily, after biogas has been switched off for one minute. Biogas burners and pipes should be cleaned every second day to avoid clogging with water vapour and ensure the flow of gas. Ensure that the valve is switched off during maintenance and clean gas holes with a small cloth. Detach the flexible pipe from the biogas pipe and clean the connection.

Table 44:
Other problems and nuisances

3 List of abbreviations

ABR	Anaerobic Baffled Reactor
AF	Anaerobic Filter
BALIFOKUS	BaliFokus Foundation (NGO), Indonesia
BAPPENAS	National Development Planning Agency, Indonesia
BEST	Bina Ekonomi Sosial Terpadu – Institute for Integrated Economic & Social Development (NGO), Indonesia
BMZ	Bundesministerium für wirtschaftliche Zusammenarbeit und Entwicklung – German Federal Ministry for Economic Cooperation and Development
BNS	Basic Needs Services
BOD	Biological Oxygen Demand
BORDA	Bremen Overseas Research and Development Association (NGO), Germany
CBS	Commuity-Based Sanitation
CDD	Consortium for DEWATS Dissemination Society (NGO), India
CERNA	Centre d'Economie Industrielle; the Cerna is the Centre of Industrial Economics at Mines ParisTech
CESR	Center for Environmental System Research, University of Kassel, Germany
COD	Chemical Oxygen Demand
CPCB	Central Pollution Control Board, under the Indian Ministry of Environment and Forests
CSC	Community Sanitation Centre
DALY	Disability Adjusted Life Year
DED	Detailed Engineering Design
DEWATS	Decentralised Wastewater Treatment System(s)
DM	Dry Matter
DO	Dissolved Oxygen
EAWAG	Swiss Federal Institute of Aquatic Science and Technology
EEA	European Environment Agency
EoI	Expression of Interest
GIS	Geographical Information System
GDP	Gross Domestic Product
GP	Gram Panchayat (Indian Government Administration on Village level)
HRT	Hydraulic Retention Time
IDR	Indonesian Rupiah (currency)
INR	Indian Rupees (currency)
IPLT	Indonesian name for municipal sludge treatment plant
JMP-WHO/ UNICEF	The Joint Monitoring Programme for Water Supply and Sanitation is co-funded by WHO and UNICEF
LCA	Life Cycle Assessment
LCM	Life Cycle Management
LPTP	Lembaga Pengembangan Teknologi Pedesaan (NGO), Indonesia
Ltd.	Private company limited by shares
MoU	Memorandum of Understanding

NGO	Non Governmental Organisation
NTU	Standardized Degree of Turbidity
OTM	Organic Total Solids
O&M	Operation and Maintenance
PE-HD	or HDPE – High Density Polyethylene
pH	pH is a measure of the acidity or basicity of a solution
PHAST	Participatory Hygiene and Sanitation Transformation
PhP	Philippines Pesos (currency)
PVC	Polyvinyl Chloride
RPA	Rapid Participatory Assessment
SANIMAS	Sanitation by Neighbourhoods, Community-Based Sanitation Project, 2001-04 financed by a trustfund from AusAID managed by WSP East Asia. Ongoing as national sanitation programme under multi-finance scheme (Ministry of Public Works, local level authorities and beneficiaries)
SDSI	Sustainable Development Strategy Institute
SENA	Chinese State Environmental Protection Administration
SME	Small and Medium Entities
SS	Suspended Solids
TOC	Total Organic Carbon
TS	Total Solids
UASB	Upflow Anaerobic Sludge Blanket
UNDP-HDR	The Human Development Report is an annual milestone publication by the United Nations Development Programme
UNICEF	United Nations Children's Fund (originally United Nations International Children's Emergency Fund)
UN-WWDR	United Nations World Water Development Report
UK	United Kingdom
US	United States
US EPA	United States Environmental Pollution Agency
UV	Ultraviolett radiation
VFA	Volatile Fatty Acids
VIP	Ventilated Improved Pit Latrine
VS	Volatile Solids
WaterGAP	Water, Global Assessment and Prognosis (Modelling Tool)
WEDC	Water, Engineering and Development Centre, Loughborough University, United Kingdom
WEFTEC	Water Environment Federation's annual Technical Exhibition and Conference
WHO	World Health Organisation
WSP	Water and Sanitation Program
WWAP	World Water Assessment Programme
ZP	Zilla Parishad (Indian Government Admistration on District level)
ZUT	Zhejiang University of Technology

4 Appendix

14.1 Geometric formulas

Geometric formulas		
rectangle	$A = a \times b$	
rectangular prism	$A = 2 \times (a \times b + a \times c + b \times c)$	$V = a \times b \times c$
trapezium	$A = \frac{a+c}{2} \times h$	
trapeziform prism		$V = \frac{h}{2} \times h \, (a \times b + c \times d + \sqrt{a \times b \times c \times d})$
circle	$A = \pi \times r^2$	$C = 2 \times \pi \times r$
cylinder	$A \text{ (mantle)} = 2 \times \pi \times r \times h$	$V = \pi \times r^2 \times h$
sphere (ball)	$A = 4 \times \pi \times r^2$	$V = \frac{4}{3} \times \pi \times r^3$
spherical segment	$A = 2 \times \pi \times r \times h$	$V = \pi \times h^2 \times (r - \frac{h}{3})$
cone	$A \text{ (mantle)} = \pi \times r \times s$	$V = \pi \times r^2 \times \frac{h}{3}$
law of pythagoras	$a^2 + b^2 = c^2$	sides of 90° triangle: 3 / 4 / 5
tangent	a / b	$\tan 45° = 1$
		$\tan 30° = 0.577$
		$\tan 60° = 1.732$

Table 45: Geometric formulas

14.2 Energy requirement and cost of pumping

A	B	C	D	E	F	G	H	I
Energy requirement and cost of pumping								
flow rate	main flow h/d	flow rate per hour	pump high	assumed head loss	efficiency of pump	required power of pump	cost of energy	annual energy cost
m³/d	h	m³/h	m.	m	η	kw	ECU/kWh	ECU
26	10	2.6	10	3	0.5	0.18	0.15	100.85

$C4 = A4 / B4$

$G4 = 9.81 \times (D4 + E4) \times C4 / F4 / 3600$

$I4 = B4 \times G4 \times 365 \times H$

Table 46: Energy requirement and cost of pumping

14.3 Sedimentation and flotation

The performance of a domestic-wastewater settler is sufficient when the effluent contains less than 0.2ml/l settleable sludge after a 2h jar test.

The general formula for calculating the surface area for floatation and sedimentation tanks is:

Water surface [m²] = water volume [m³/h] /
slowest settling (floatation) velocity [m/h].

Settling and floatation velocity can be calculated by observing the settling process in a glass cylinder. The formula is:

Settling (floatation) velocity [m/h] =
height of cylinder [m] / settling (floatation) time [h]

Flocculent sludge has a settling velocity between 0.5 and 3 m/h.
The velocity in a sand trap should not exceed 0.3 m/s (1000 m/h).
The minimum cross section area is then:

Area [m²] = flow [m³/s] / 0.3 [m/s], or
Area [m²] = flow [m³/h] / 1000 [m/h]

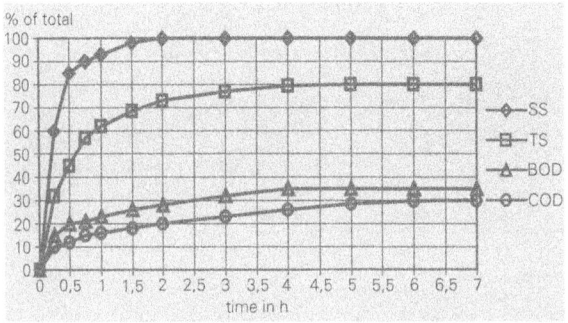

Appendix_1:
Removal rates in settling tests of domestic wastewater

The above graph shows the results of settling tests in a jar test under batch conditions (SS = settleable solids, TS = total solids; COD is measured as COD_{KMnO_4}). The curve might be different in through-flow settlers. The more turbulent the flow, the lesser the removal rate of settleable solids; however, BOD- and COD-removal rates increase with more complete mixing of old and new wastewater.

14.4 Flow in partly filled round pipes

A	B	C	D	E	F	G	H	I	J
\multicolumn{10}{c}{Flow in partly filled round pipes}									
pipe	flow height	flow area	moisted area/m	hydraulic radius	slope	roughness	flow speed	flow	flow
chosen	given	calcul.	calcul.	calcul.	chosen	estimat.	calcul.	calcul.	calcul.
d	h/d	A	U	rhy	s	rf	v	Q	Q
m	m/m	m²	m	m	%		m/s	l/s	m³/h
0.1	0.15	0.00074	0.080	0.0093	1.0%	0.35	0.21	0.153	0.55
0.1	0.25	0.00154	0.105	0.0147	1.0%	0.35	0.31	0.478	1.72
0.1	0.35	0.00245	0.127	0.0194	1.0%	0.35	0.40	0.969	3.49
0.1	0.50	0.00393	0.157	0.0250	1.0%	0.35	0.49	1.932	6.96
0.1	0.75	0.00632	0.210	0.0302	1.0%	0.35	0.58	3.641	13.11

Table 47:
Flow in partly filled round pipes

Formulas of spreadsheet for "flow in partly filled pipes" (after Kutter's short formula)

$$C6 = 0.295 \times (A6/2)^2$$

All figures – as here 0.295 – are geometrical constants, referring to the flow height in relation to the diameter of the pipe.

$D6 = 1.591 \times (A6 / 2)$

$E6 = C6 / D6$

$H6 = (100 \times \sqrt{E6} / (G6 + \sqrt{E6})) \times \sqrt{E6 \times F6}$

$I6 = C6 \times H6 \times 1000$

$J6 = I6 \times 3.6$

$C7 = 0.614 \times (A7 / 2)^2$

$D7 = 2.094 \times (A7 / 2)$

$E7 = C7 / D7$

$H7 = (100 \times \sqrt{E7} / (G7 + \sqrt{E7})) \times \sqrt{E7 \times F7}$

$I7 = C7 \times H7 \times 1000$

$J7 = I7 \times 3.6$

$C8 = 0.98 \times (A8 / 2)^2$

$D8 = 2.532 \times (A8 / 2)$

$E8 = C8 / D8$

$H8 = (100 \times \sqrt{E8} / (G8 + \sqrt{E8})) \times \sqrt{E8 \times F8}$

$I8 = C8 \times H8 \times 1000$

$J8 = I8 \times 3.6$

$C9 = 1.571 \times (A9 / 2)^2$

$D9 = 3.142 \times (A9 / 2)$

$E9 = C9 / D9$

$H9 = (100 \times \sqrt{E9} / (G9 + \sqrt{E9})) \times \sqrt{E9 \times F9}$

I9 = C9 x H9 x 1000
J9 = I9 x 3.6
C10 = 2.528 x (A10 / 2) ^ 2
D10 = 4.19 x (A10 / 2)

E10 = C10 / D10
H10 = (100 x SQRT (E10) / (G10 + SQRT (E10))) x SQRT (E10 x F10)
I10 = C10 x H10 x 1000
J10 = I10 x 3.6

14.5 Conversion factors of US-units

	Conversion factors of US-units			
item	US-unit	SI-unit	US/SI-unit	SI/US-unit
length	in.	cm (10mm)	2.540	0.394
	ft (12in)	m (100mm)	0.305	3.281
	yd (3ft)	m	0.914	1.094
	mi (1,760yd)	km (1,000m)	1.609	0.621
area	in^2	cm^2	6.452	0.155
	ft^2	m^2	0.093	10.764
	yd^2	m^2	0.836	1.196
	acre	hectar (10,000m^2)	0.405	2.471
	mi^2	km^2	2.590	0.386
volume	in^3	cm^3	16.387	0.061
	ft^3	liter	28.317	0.035
	ft^3	m^3	0.0283	35.314
	gallon	litre	3.785	0.264
	yd^3 (202gal)	m^3	0.765	1.308
	acre-foot	m^3	1,233.5	0.811
force / mass	lb	N	4.448	0.225
	oz	g	28.350	0.035
	lb (16oz)	kg (1,000kg)	0.454	2.205
	ton (short) (2,000lb)	t (1,000kg)	0.907	1.102
	ton (long) (2,240lb)	t (1,000kg)	1.016	0.984
pressure	in H_2O	Pa (N/m^2)	204.86	0.005
	lb/in^2	kPa (kN/m^2)	6.895	0.145
	lb/in^2	Pa (N/m^2)	47.88	0.021
flow rate	gal/min	l/s (86.4m^3/d)	0.0631	15.850
	gal/d	l/s	0.0000438	22,825
	gal/min (1,440gal/d)	m^3/d (0.0116l/s)	0.00379	264
energy + power	Btu	kJ	1.055	0.948
	hp-h	MJ	2.685	0.373
	kWh	kJ	3,600	0.00028
	Ws	J	1,000	0.001
	hp	kW	0.746	1.341
temperature	°F	°C	0.56(°F-32)	1.8(°C)+32
	°F	K	0.56(°F+460)	1.8(°K)-460

Table 48: Conversion factors of US-units

5 Bibliography

Abwassertechnische Vereinigung e.V. (ed.), *"Lehr- und Handbuch der Abwassertechnik"*, Vol. I-VI, 3rd Edition, Wilhelm Ernst und Sohn, Berlin, München, Düsseldorf, 1982

ADB (Asian Development Bank), *"Water in Asian Cities. Utility Profile."* Manila, 2006

Alaerts, G.J., Veenstra, S., Bentvelsen, M., van Duijl, L.A. at al., "Feasibility of Anaerobic Sewage Treatment in Sanitation Strategies in Developing Countries" IHE Report Series 20, International Institute for Hydraulic and Environmental Engineering, Delft, 1990

Allen, Adriana, Julio Davila, and Pascale Hoffman. *"Governance of Water and Sanitation Services for the Peri-Urban Poor: A Framework for Understanding and Action in Metropolitan Regions."* University College London, Development Planning Unit, London, 2006

Bachmann, A., Beard, V.L. and McCarty, P.L., *"Performance Characteristics of the Anaerobic Baffled Reactor"*, Water Science and Technology, IAWQ, 1985

Bahlo, K., Wach, G., *"Naturnahe Abwasserreinigung – Planung und Bau von Pflanzenkläranlagen"*, 2nd Edition, Ökobuch, Staufen bei Freiburg, 1993

Batchelor, A. and Loots, P., *"A Critical Evaluation of a Pilot Scale Subsurface Flow Wetland: 10 Years After Commissioning"*, Water Science and Technology, Vol. 35, No. 5, IAWQ, 1997

Barber, W.P., Stuckey and D.C., *"The Use of the Anaerobic Baffled Reactor (ABR) for Wastewater Treatment: a review"*, Water Research, 1999

Bischof, W. (Hosang/Bischof), *"Abwassertechnik"*, 9th edit., B.G. Teubner Verlag, Stuttgart, 1989

Blum, D. and Feachem, R.G., *"Health Aspects of Nightsoil and Sludge Use in Agriculture and Aquaculture"* IRCWD Report No. 05/85, (reprint 1997), EAWAG / SANDEC, Dübendorf CH, 1997

Boller, M., *"Small Wastewater Treatment Plants – a Challenge to Wastewater Engineers"*, Water Science and Technology, Vol. 35, No. 6, IAWQ, 1997

Brix, H., *"Do Macrophytes Play a Role in Constructed Wetlands?"*, Water Science and Technology, Vol. 35, No. 5, IAWQ, 1997

California Energy Commission, *"California's Water-Energy Relationship"*, Final Staff Report, CEC-700-2005-011-SF, Sacramento, 2005

Central Pollution Control Board, *"Guidelines on Environmental Management in Industrial Estates"*, Programme Objective Series, Probes/43/1989-90, Delhi, 1990

Central Pollution Control Board, *"Sewage Pollution"*, Parivesh Newsletter, Delhi, 2005

Cooper P.F., Hobson J.A., Findlater C., *"The Use of Reed Bed Treatment Systems in the UK"*, Proceedings of conference on small wastewater treatment plants, edited by Hallvard Odegaard, Trondheim, 1989

Cooper, P.F (ed.), *"European Design and Operations Guidelines for Reed Bed Treatment Systems"* (revised document), EC/EWPCA Emergent Hydrophyte Treatment Systems Expert Contact Group, Swindon, UK 1990

Cross, P., Strauss, M., *"Health Aspects of Nightsoil and Sludge Use in Agriculture and Aquaculture"*, IRCWD, Dübendorf CH, 1986

De Vries J., *"Soil filtration of wastewater effluent and mechanism of pore . clogging"*, JWPCF Journal of Water Pollution Control Federation, vol.44, 1972

Denny, P., *"Implementation of Constructed Wetlands in Developing Countries"*, 5th International Conference on Wetland Systems for Waste Pollution Control, Vienna, 1996

Department of Environmental Protection and Energy, Ministry of Agriculture, P.R. of China, *"Biogas and Sustainable Agriculture"*, Collection of Papers from the National Experience Meeting in Yichang, Hubei Province 1992, published in co-operation with BORDA, Bremen, 1993

5 Bibliography

Driouache, A. et al., *"Biogasnutzung in der Abwasserstation von Ben Sergao (Marokko) – Methoden und Ergebnisse"*, CDER/PSE (GTZ), Eschborn, 1997

Driouache, A. et al., *"Promotion de l'utilisation du biogaz produit dans des stations d'epuration au Maroc"*, CDER/PSE (GTZ), Marrakech, 1997

Ellenberg, H. et al., *"Biological Monitoring – Signals from the Environment"*, GATE/GTZ publication, Vieweg Verlag, Eschborn/ Braunschweig, 1991

European Environment Agency, *EEA Glossary*, Copenhagen, 2006

FAO (Food and Agriculture Organization of the United Nations) and Jelle Bruinsma (ed.). World Agriculture: Towards 2015/2030 – An FAO Perspective, London, 2003

Fasteneau F., Graaf J. and Martijnse G., *"Comparison of Various Systems for On-Site Wastewater Treatment"*, Proceedings of Conference on Small Wastewater Treatment Plants, edited by Hallvard Odegaard, Trondheim, 1989

Friedrich E., Pillay S.D and Buckley C.A., *"The Use of Life Cycle Assessment (LCA) in the Water Industry and the Case for an Environmental Performance Indicator"*, Water SA, 33 (4), 143-159, 2007

Friedrich E., Pillay S.D and Buckley C.A, *"Environmental Life Cycle Assessment – A South African Case Study of an Urban Water Cycle"*, Water SA, 35 (1), 73-84, 2009

Friedrich, E. and Buckley, C.A., *"The Application of LCA within LCM for Municipal Water Systems – Are We Closing the Loop?"*, Proceedings of the 4th International Conference on Life Cycle Management, Cape Town, 2009

Garg, S. K., *"Sewage Disposal and Air Pollution Engineering"*, 9th revised edit., Khanna Publishers, New Delhi, 1994

Geller, G., *"Horizontal Subsurface Flow Systems in the German Speaking Countries: Summary of Long-Term Scientific and Practical Experiences"*, Water Science and Technology, Vol. 35, No. 55, IAWQ, 1997

Grau, P., *"Low Cost Wastewater Treatment"*, Water Science and Technology, Vol. 33, No 8, IAWQ, 1996

GRET, *"Water and Health in Underprivileged Urban Areas"*, PS Eau – Edition GRET, France, 1994

Global Water Partnership, *"Desafíos de la regulación de los servicios de agua y saneamiento en América Latina"*, Stockholm, 2004

Government of South Africa, *"White Paper on Basic Household Sanitation"*, Pretoria, 2001

Gutterer, B., *"Decentralized Waste Water Treatment in Developing Countries – an E-learning course"*, Bremen/Berlin, 2004

Gutterer, B., *"DEWATS in India – an Evaluation Report"*, Berlin, 2007

Grobicki, A. and Stuckey, D.C., *"Performance of the Anaerobic Baffled Reactor"*, Poster Papers, Fifth International Symposium on Anaerobic Digestion, Bologna, May 1988

GTZ, Grundlagen *"Verbesserung der Entsorgung in städtischen Armutsgebieten"*, Eschborn, 2002

GTZ/Ecosan, *"Decision-Making Tools Papers of the 2nd Ecosan Symposium"*, Lübeck, Germany, 2003

Cities in a Globalizing World. *"Global Report on Human Settlements"*, London, 2001

Hagendorf, U., *"Verbleib von Abwasserinhaltsstoffen bei Pflanzenkläranlagen im Langzeitbetrieb"*, FGU-Seminar: Naturnahe Abwasserbehandlung durch Pflanzenkläranlagen, Berlin, 1997

Hamburger Umwelt Institut e. V. and O Instituto Ambiental, *"Biomass Nutrient Recycling – purifying water through agricultural production"*, Hamburg/Silva Jardim, 1994

6 Bibliography

Hanqing Yu, Joo-Hwa Tay and Wilson, F., *"A Sustainable Municipal Wastewater Treatment Process for Tropical and Subtropical Regions in Developing Countries"*, Water Science and Technology, Vol. 35, No. 9, IAWQ, 1997

Heinss, U., Larmie, S.A., Strauss, M., *"Solids Separation and Pond Systems for the Treatment of Faecal Sludges in the Tropics"*, EAWAG/SANDEC, Dübendorf CH, 1998

Imhoff, K. and Imhoff, K.R., *"Taschenbuch der Stadtentwässerung"*, 27th edit, R. Oldenbourg Verlag, München, Wien, 1990

Inamori, Y. et.al., *"Sludge Production Characteristics of Small Scale Wastewater Treatment Facilities Using Anaerobic/Aerobic Biofilm Reactors"*, Water Science and Technology, Volume 34, No. 3-4, IAWQ, 1996

IWMI (International Water Management Institute). *"Recycling Realities: Managing Health Risks to Make Wastewater an Asset."* Water Policy Briefing 17. Colombo, 2006

Jenssen, P.D. and Siegrist, R.L., *"Technology Assessment of Wastewater Treatment by Soil Infiltration Systems"*, Proceedings of Conference on Small Wastewater Treatment Plants, edited by Hallvard Odegaard, Trondheim, 1989

Johnstone, D.M.W. and Horan, N.J., *"Institutional Development Standards and River Quality: A UK History and Some Lessons for Industrialising Countries"*, Water Science and Technology, Vol. 33, No. 3, IAWQ ,1996

Kadlec, Robert H. and Knight, Robert L., *"Treatment Wetlands"*, Lewis Publishers; Boca Raton, New York, London, Tokyo, 1996

Karstens, A. and Berthe-Corti, L., *"Biologischer Hintergrund zur anaeroben Klärung von Abwässern"* University of Oldenburg/BORDA,1996

Kazaglis, A., Kraemer, P., *"Sanitation Success Stories in India and Implications for Urban Sanitation Planning"*, 32nd WEDC International Conference, Colombo, 2006

Khanna, P. and Kaul, S.N. *"Appropriate Waste Management Technologies for Developing Countries"*, Selected proceedings of the 3rd IAWQ spezialized conference on Appropriate Waste Management Technologies for Developing Countries, Nagpur, 1995

Knoch, W., *"Wasserversorgung, Abwasserreinigung und Abfallentsorgung – Chemische und analytische Grundlagen"*, VCH Verlagsgesellschaft, Weinheim, 1991

Kraemer, P., Ulrich, A., *"Mainstreaming Community Based Sanitation in Urban Areas of South Asia"*, 32nd WEDC International Conference, Colombo, 2006

Kraft, H., *"Alternative Decentralized Sewage Systems in Regions of Dense Population – Root Zone Treatment Plants and other Systems"*, Indo-German Workshop on Root Zone Treatment Systems, CPCB/GTZ, New Delhi, 1995

Krekeler, T., *„Decentralised Sanitation and Wastewater Treatment"*, revised 2nd edition, Bundesanstalt für Geowissenschaften und Rohstoffe (BGR), Hannover, 2008

Kunst, S. and Flasche, K., *"Untersuchung zur Betriebssicherheit und Reinigungsleistung von Kleinkläranlagen mit besonderer Berücksichtigung der bewachsenen Bodenfilter"*, Abschlussbericht, ISAH, University of Hannover, 1995

Lago, S. and Boutin, C., *"Etude comparative des procédés d'épuration biologique dans le contexte africain"*, CEMAGREF L132, 1991

Liao Xianming, *"Anaerobic Digester for Stabilizing Domestic Sewage"*, BRTC, Chengdu, 1993

Lienard, A. and Coll., *"Coupling of Reed Bed Filters and Ponds – an example in France"*, CEMAGREF, Lyon, 1994

Loll, U., *"Vergleich zwischen Wurzelraumentsorgungs- und Teichkläranlage am Projekt Brandau"* BMFT/ATV Seminar: Naturnahe Verfahren der Abwasserbehandlung, Pflanzenkläranlagen, Essen-Heidhausen, 1989

Lusaka Water and Sewerage Company "Urban Development Plans and Infrastructure for the City of Lusaka", Lusaka, 2005

Maeseneer, J.D. and Cooper, P., *"Vertical Versus Horizontal-Flow Reedbeds for Treatment of Domestic and Similar Wastewaters"*, IAWQ Specialist Group Newsletter No. 16, June 1997

Mara, D., *"Design Manual for Stabilization Ponds in India"*, Ministry of Environment and Forests/DFID, published by Lagoon Technology International Ltd., Leeds, 1997

McCarty, P. L., *"Towards Sustainability – A Paradigm Shift in Concepts, Analyses, and Goals"*, keynote-presentation at WEFTEC, San Diego, 2007

Metcalf and Eddy: *"Wastewater Engineering, Treatment – Disposal – Reuse"*, 3rd Edition, TATA McGraw-Hill Edition, New Delhi, 1996

Morgan, P., *"Toilets That Make Compost"*, Practical Action Publishing, Stockholm, 2008

Mudrak, K. and Kunst, S., *"Biologie der Abwasserreinigung"*, 3rd Edition, Gustav Fischer Verlag, Stuttgart, 1991

National Development Planning Agency/Bappenas: National Policy: *"Development of Community Managed Water Supply And Environmental"*, Jakarta, 2002

ODI (Overseas Development Institute). *"Right to Water: Legal Forms, Political Channels"*. ODI Briefing Paper, London, 2004

Orozco, A., *"Anaerobic Wastewater Treatment Using an Open Plug Flow Baffled Reactor at Low Temperature"*, Poster Papers, Fifth International Symposium on Anaerobic Digestion, Bologna, May 1988

Pande, D.R., De Poli, F. and Tilche, A., *"Some Aspects of Horizontal Design Biogas Plants"* Poster Papers, Fifth International Symposium on Anaerobic Digestion, Bologna, May 1988

Pandey, G.N. and Carney, G.C., *"Environmental Engineering"*, Tata McGraw-Hill Publishing Company Ltd, Delhi, 1992

Pedersen, P., *"Ecology and Urban Planning"*, Working Paper No.1, Dept. Of Human Settlements, The Royal Danish Academy of Fine Arts, Copenhagen, 1991

Platzer, Chr. and Mauch, K., *"Evaluations Concerning Soil Clogging in Vertical Flow Reed Beds"*, 5th International Conference on Wetland Systems for Waste Pollution Control, Vienna, 1996

Polprasert, C., *"Organic Waste Recycling"*, Asian Insitute of Technology, Bangkok, 1989

Reuter, S., Wendland, C., Samwel, M., Bender, M. et.al., *„Safe Sanitation: A Challenge We Can Meet Together – Policy Paper on Sustainable Sanitation"*, German NGO-Forum on Environment and Development, 2008

Reuter, S., Wendland, C., von Münch, E., Hänel, M., Sundberg, C., Platzer, C. et.al., *„Links Between Sanitation, Climate Change and Renewable Energies"*, SuSanA Factsheet WG03, Eschborn, 2009

Rybczynski, W., Polprasert, C.M., and McGarry, *"Low Cost Technology Options for Sanitation – A State of the Art Review and Annotated Bibliography"*, IDRC 102 e, Ottawa, 1978

Sasse, L. and Otterpohl, R., *"Status Report on Decentralised Low Maintenance Wastewater Treatment Systems (LOMWATS)"*, produced by BORDA in co-operation with AFPRO, CEEIC, GERES, HRIEE and SIITRAT, Bremen, 1996

Sasse, L., Kellner, C. and Kimaro, A., *"Improved Biogas Unit for Developing Countries"*, GATE Publication, Vieweg Verlag, Braunschweig, Wiesbaden, 1991

Schertenleib, R., *"Improved Traditional Nightsoil Disposal in China – An Alternative to the Conventional Sewerage System?"*, SANDEC-News, No.1, Dübendorf CH, 1995

6 Bibliography

Schierup, H.H. and Brix, H., *"Danish Experience with Emergent Hydrophyte Treatment Systems (EHTS) and Prospects in the Light of Future Requirements to Outlet Water Quality"*, Proceedings of Conference on Small Wastewater Treatment Plants, edit. Hallvard Odegaard, Trondheim, 1989

Shilton, Andy N., Prasad Julius N., *"Tracer Studies of a Gravel Bed Wetland"*; in: *"Water Science and Technology"*, Vol. 34, No. 3-4, pp 421-425, IWA, London, 1996

Shuval, H., Coll, J., *"Wastewater Irrigation in Developing Countries – Health Effects and Technical Solutions"*, World Bank No. 51, Washington DC, 1986

Strauss, M., Blumenthal, U.J., *"Human Waste Use in Agriculture and Aquaculture"* Executive Summary, IRCWD Report No. 09/90, Dübendorf CH, 1990

Suwarnarat, K., *"Abwasserteichverfahren als Beispiel naturnaher Abwasserbehandlungsmaßnahmen in tropischen Entwicklungsländern"*, (Dissertation), Darmstadt, 1979

Tilley, E., Luethi, C., Morel, A., Zurbrügg, C., Schertenleib, R., *"Compendium of Sanitation Systems and Technologies"*, EAWAG-Sandec, Dübendorf, 2008

Ulrich, A., Reuter, S. *"Community Based Sanitation Program in Tangerang and Surabaya/Indonesia"* in: *"Proceedings of the 2nd International Symposium on Ecological Sanitation"*, GTZ and IWA, Eschborn 2004

Ulrich, A., Thomas, A. *"Mainstreaming Community Based Sanitation in Urban Areas of South East Asia"*, 30th WEDC International Conference, Vientiane, 2004

UNDP, *"Human Development Report"*, New York, 2006

UNEP, *"Global Environment Outlook 3"*, Nairobi, 2002

UNEP, *"Global Environmental Outlook 4"*, Nairobi, 2007

UNESCO, World Water Assessment Programme (WWAP), *"UN World Water Development Report"*, Paris, 2003

UNESCO, WWAP, *"UN World Water Development Report"*, 2nd Edition, Paris, 2005

UNESCO, WWAP, *"UN World Water Development Report"*, 3rd Edition, Paris, 2009

United States Environmental Protection Agency (EPA), *"Constructed Wetlands and Aquatic Plant Systems for Municipal Wastewater Treatment"*, EPA/625/1-88/022, Cincinnati, 1988

United States Environmental Protection Agency (EPA), *"Small Community Wastewater Systems"*, EPA/600/M-91/032, Washington, 1991

United States Environmental Protection Agency (EPA), *"Wastewater Treatment / Disposal for Small Communities"*, Design Manual, EPA/625/R-92/005, Cincinati, 1992

United States Environmental Protection Agency (EPA), *"Global Mitigation of Non-CO_2 Greenhouse Gases"*, EPA 430-R-06-005, Washington, 2006

Water and Sanitation Program East Asia and the Pacific (WSP-EAP), *"Philippines Sanitation Sourcebook and Decision Aid"*, Jakarta, 2005

Wendland, C., *"Anaerobic Digestion of Blackwater and Kitchen Refuse"*, PhD thesis, Hamburg University of Technology, Hamburg, 2009

WHO, *"WHO Guidelines for the Safe Use of Wastewater, Excreta and Greywater, Volume I, Policy and Regulatory Aspects, Volume II, Wastewater Use in Agriculture, Volume III, Wastewater and Excreta Use in Aquaculture, Volume IV, Safe Use of Wastewater, Excreta And Greywater"*, Geneva, 2006

WHO/UNICEF, JMP, *"Meeting the MDG Drinking Water and Sanitation Target: A Mid-Term Assessment of Progress"*, Geneva, 2004

WHO/UNICEF, JMP, *"Water for Life: Making it Happen"*, Geneva, 2005

WHO/UNICEF, JMP, *"Meeting the MDG Drinking Water and Sanitation Target: the Urban and Rural Challenge of the Decade"*, Geneva, 2006

6 Bibliography

WHO/UNICEF, JMP, *"Progress on Drinking Water and Sanitation; Special Focus on Sanitation"*, Geneva, 2008

Winblad, Uno Ed: *"Ecological Sanitation"*, Sida, Stockholm, 1998

Winblad, U. and Kilama, W., *"Sanitation Without Water"*, revised and enlarged Edition, Macmillan, London 1985

Wissing F., *"Wasserreinigung mit Pflanzen"*, Verlag Eugen Ulmer, Stuttgart, 1995

Werner, C. "Ecosan Speech and Presentation at UN-CSD 12", New York, 2004

World Bank and World Health Organization, *"Health Aspects of Wastewater and Excreta Use in Agriculture and Aquaculture – The Engelberg Report"*, IRCWD, No. 23, 1985

World Bank, *"Environment Monitor Thailand and Philippines"*, Manila, 2005

World Bank, *"Philippines Environment Monitor"*, 2003

World Bank, *"Indonesia Environment Monitor"*, Jakarta, 2003

World Commission on Water, *"Water for Life"*, Delft, 1997

World Health Organization, *"Health Guidelines for the Use of Wastewater in Agriculture and Aquaculture"*, WHO Technical Report Series 778, Geneva, 1989

World Health Organization /UNICEF, *"Global Water Supply and Sanitation Assessment, 2000 Report"*, Geneva, 2000

Yang Xiushan, Garuti, G., Farina, R., Parsi; V. and Tilche, A., *"Process Differences Between a Sludge Bed Filter and an Anaerobic Baffled Reactor Treating Soluble Wastes"*, Poster Papers, at the Fifth International Symposium on Anaerobic Digestion, Bologna, May 198

Photos:
BORDA Network Archive

Photos courtesy of:
Hesperian Foundation: p.298
Peter Morgan: p. 292
Inge Maltz: p.318
Roediger Vakuum- und Haustechnik GmbH: p. 308, p. 309

Graphics courtesy of
SANIMAS: p. 59, p. 93, p. 123, p. 289, p. 297, p. 300, p. 301, p. 302, p. 304

Graphic Design, Layout:
www.conrat.org

www.ingramcontent.com/pod-product-compliance
Ingram Content Group UK Ltd.
Pitfield, Milton Keynes, MK11 3LW, UK
UKHW021839210426
5322IPUK00021B/365